Introduction to Crystal and Electronic Structures of Solids

- Yoon, Sokkeel -

UUP

Introduction to Crystal and Electronic Structures of Solids

2007년 1월 15일 초판 인쇄
2007년 1월 20일 초판 발행

지은이 | 윤석길
발행인 | 정정길
발행처 | UUP(울산대학교 출판부)

출판등록 | 제 370-1996-000001호(1996.3.13)
주소 | 680-749 울산광역시 남구 무거 2동 산29번지
전화 | (052) 259-2488
팩스 | (052) 277-3011
편집·표지 디자인 | 지미정·김수연

값 5,000원
ISBN 978-89-7868-956-4

Preface

How to give various physical concepts about condensed matters to students who are lack of advanced mathematical tools is the purpose of writing this book. Basic concepts on atomic bonding, crystal structures, lattice and wave, lattice vibrations and thermal properties, free electrons and energy bands, were dealt with least mathematics and short explanations as plain as possible. I hope this book does a good role as a pre-reading for students before starting studies on the electromagnetic properties of solids.

Comments on any mistakes in contents as well as English will always be appreciated.

November 2006

Yoon, Sokkeel

Contents

Preface / 3

Chapter One Atomic Bonding in Solids 9

1.1 Types of Atomic Bonding 9
1.1.1 Ionic bonding 9
1.1.2 Covalent bonding 11
1.1.3 Metallic bonding 13
1.1.4 Molecular bonding-van-de-Waal's bonding · 14
1.1.5 Hydrogen bonding 15
1.2 Equilibrium Inter-atomic Distance 16

Chapter Two Crystal Structure 19

2.1 Crystal 19
2.1.1 Lattice and symmetry 19

2.1.2 Translational symmetry operation and unit cell ··· 20

2.1.3 Point group symmetry operation ··· 21

2.1.4 Space group ··· 23

2.1.5 Primitive unit cell, centered unit cell and 14 Bravais lattices ··· 24

2.2 Simple Crystal Structures ··· 32

2.3 Wave and crystal ··· 40

2.3.1 Reciprocal lattice ··· 41

2.3.2 Brillouin zone ··· 44

Chapter Three Lattice Wave and Specific heat of Solids ···· 51

3.1 Harmonic Oscillation of Atoms in Solids ··· 51

3.1.1 One dimensional homogeneous line ··· 53

3.1.2 One dimensional monatomic lattice ··· 58

3.1.3 One-Dimensional Diatomic lattice ··· 63

3.2 Quantization of Lattice Waves ··· 67

3.3 Specific Heat of Solid ··· 71

3.3.1 Dulong-Petit's law ··· 71

3.3.2 Theory of specific heat of solids ··· 73

3.3.3 Phonon and zero point energy ··· 85

3.3.4 Thermal conductivity ··· 86

3.3.5 Thermal expansion ··· 92

Chapter Four Electrons in Metal - Energy Band Theory 97

4.1 Classical Theory of Free Electron 97
4.1.1 Electric conduction - Ohm's law 98
4.1.2 Thermal conduction 101
4.1.3 Failures of the classical free electron model . 103
4.2 Quantum Theory of Free Electron 105
4.2.1 Quantization of electronic states 105
4.2.2 Fermi energy and Fermi surface 109
4.2.3 Density of states 112
4.2.4 Failures of the simple quantum theory of free electron 117
4.3 Energy Band Theory 121
4.3.1 Nearly free electron model 121
4.3.2 Density of states 126
4.3.3 Metal, insulator and semiconductor 129
4.3.4 Effective mass 132
4.3.5 Holes 136
4.3.6 Linear combination of atomic orbitals 138

Chapter One

Atomic Bonding in Solids

1.1 Types of Atomic Bonding

1.1.1 Ionic bonding

- When an atom with loosely bound outermost electron(s) (atom with high ionization tendency, Na, for instance) meets an atom with tightly bound vacant seat(s) for electron(atom with high electron affinity, Cl, for instance), the loosely-bound electron(s) happily transfers to the other atom to be in more tightly bound state. This process lowers total energy of the system of the two atoms. As a result, a positive ion(Na^+, say) and a negative ion(Cl^-) appear.
- Two ions of opposite sign attract each other via electro-

static force. The electro-static force(Coulomb force) between two ions is direct, strong and isotropic, so that ionic crystals have high melting points and close pact type structures. Since this type of bonding is sharply dependent on the inter-ionic distance($\propto 1/r^2$), ionic crystals are brittle and poor in machinability.

- All the electrons in ionic crystals are tightly bound by ions leaving no free electrons. Without charge carriers, ionic materials are insulators electrically and also transparent to the electromagnetic waves since there is no particles absorbing the energy of incident photons.

- At elevated temperatures, however, some electrons absorb enough energy to escape from the orbital and become 'free'. This explains the reason why ionic crystals become (semi)conductive and opaque at high temperatures. Similarly, electromagnetic waves of higher frequencies (photons with higher energy, $E=h\nu$) can knock the orbital electrons off and are absorbed in the ionic materials. In general, ionic crystals are opaque to e.m. waves of U.V. range and upwards.

1.1.2 Covalent bonding

- If an electron belongs to two 'nuclei', its energy will decrease since nucleus is positive electrically. When two atoms with similar ionization tendencies meet closer, instead one of them sends electron(s) to the other as in the ionic bonding, they invest one or two(three sometimes) electrons each and co-posses them in the form of 'molecular orbitals'. Lowering the energy as two atoms approach means an attractive force between the two atoms. This type of bonding is called the covalent bonding.

- There is two basic conditions for the covalent bonding to occur. It is natural that the molecular orbital should not be of the upper energy state than the initial orbitals. Otherwise, net energy change becomes positive and repulsive rather than attractive force will make the bonding impossible. The first condition is that the participating atoms should have vacant states in the outermost orbit. The second condition is that lobes of partially filled orbitals of each atom should overlap with each other to form the molecular orbits. This implies the covalent bonding is anisotropic.

- In many atoms, p-orbits are outermost and partially filled. p-states have higher energies than those of s-states in an isolate atom. But when the atom approaches to other atoms, it is influenced by the electric field and energies of p-state and s-state become more or less same level. As a result, there appears a sp-hybridized orbit which have four major lobes and accommodates 2 electrons at each lobe up to 8 electrons at total.

- When the atom has only two or three vacancies(two in oxygen atom), covalent bonding between two or three atoms may form a molecular orbit to fill all the vacancies by overlapping two or three lobes(like O=O). This type of covalent bonding is called the saturated covalent bonding and does not provide inter-molecular attraction.

- In some materials(e.g., Si), four lobes have one vacancy each and give a type of unsaturated covalent bonding which extends into space and form a solid.

- Covalent solids are insulator since there is no free electron. Comparing with ionic solids, however, electrons in the molecular orbit have somewhat mobile character and show often conductivity in the semiconducting range. Being optically opaque up to infra-red range in many covalent solids can also be understood by the absence of free electrons and by existence of less-tightly-bound ‘molecular’ electrons.

- Bonding via orbital lobes gives directionality to the bonding. The anisotropy of bonding results in less-compact crystal structures (small density in general) and poor machinability.

1.1.3 Metallic bonding

- When an atom with loosely-bound outermost electron(s) meets another atom of same kind, the outermost electrons do not transfer from one atom to the other like in ionic bonding since electronic situation is not different for the two atoms. Instead, the electrons in the overlapping orbitals form new ’molecular’ orbitals which belong to the two positive ions. If three atoms are involved, then the 'molecular' orbitals covers all the three ions. And so on. Positive ions are bound indirectly via negatively-charged electrons in the 'molecular' orbitals.

- We usually do not use the word 'molecular orbitals' in this case and the word 'free electrons' is commonly used instead. The metallic bonding is, therefore, often expressed as; "positive ions are immersed in the sea of free electrons". The existence and isotropic distribution of free electrons characterize the properties of metallic materials.

- Good electrical conductivity, good thermal conductivity and good absorption of electro-magnetic waves and so good reflectivity are among common properties of metals due to free electrons. Indirect and isotropic natures of the bonding gives good machinability and compact crystal structures.

1.1.4 Molecular bonding-van de Waal's bonding

- As mentioned above, molecules with saturated covalent bonding or inert atoms have negligible attraction between them. That is why most of them exist in gaseous or liquid state at room temperatures.

- Though electronic distribution in molecular or atomic orbits is homogeneous or symmetrical when time-averaged, instantaneously it is not either homogeneous or symmetrical so that individual molecule(atom) form a electric dipole fluctuating every instant. Dipole-dipole interaction is far weaker than force between point charges. Molecular bonding is based on such weak interaction between fluctuating dipoles. It may be a little bit enhanced when the fluctuation becomes synchronized. That is why molecular solids are rather soft and have very low melting points.

- Without free electrons, molecular solids are electrically insulator and optically transparent up to visible range sometimes.

1.1.5 Hydrogen bonding

- When covalent or molecular materials have hydrogen as a constituent(H_2O for example), hydrogen atoms donate their only one electron to the molecular orbit. Becoming a 'naked' proton, the hydrogen atom exhibits a very strong positivity leaving the counterpart atom relatively negative. As a result, the molecule behaves like a quasi-ionic molecule. Bonding between the hydrogen and the 'negative' atom is the base of hydrogen bonding which is somewhat stronger than molecular bonding but far weaker than ionic bonding.

- Electrical and optical properties are similar to those of molecular materials.

1.2 Equilibrium Inter-atomic Distance

- Attraction between atoms in solids is basically of electrostatic origin and can be represented by a potential, $-\frac{A}{r^2}$. On the other hand, repulsion is of quantum mechanical origin due to overlapping of inner orbitals, which may be expressed by a potential, $+\frac{B}{r^n}$, (n~10).

- Total potential energy, U, shows a minimum at $r=r_0$ and

r_0 is the equilibrium interatomic distance. $U(r_0)$ is the measure of the binding energy.(Fig 1.1)

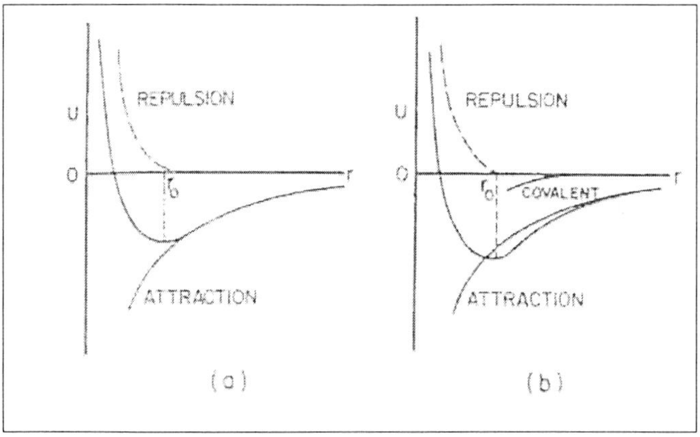

Fig 1.1 Potential energy v.s. inter-atomic distance

- If we put the origin at (r_0, $U(r_0)$), it is seen that U-x curve is parabolic when x is small, that is, $U \propto x^2$. Here, $x=r-r_0$, deviation from the equilibrium position in other words. This means that, being off the equilibrium position, an atom is exerted a force to take it back to the equilibrium position. The restoring force is linearly proportional to the deviation, x, which is just same as the Hooke's law in a spring. The range of x where this rule is obeyed is called the Hooke's law region.

$$U = \frac{1}{2}\alpha x^2$$

and

$$F = -\frac{\partial U}{\partial x} = -\alpha x.$$

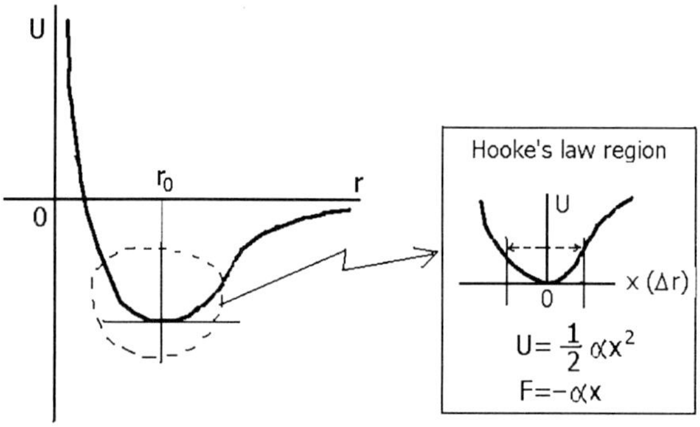

Fig 1.2 Hooke's law region

- When x becomes larger, the Hooke's law is no longer obeyed.

Chapter Two

Crystal Structure

2.1 Crystal

2.1.1 Lattice and symmetry

- Crystal means a solid state where atoms or pattern of atoms are arrayed regularly, repeatedly and periodically, to form the solid. The regularity of array is often expressed as symmetry. We can say that a crystal is a solid where atoms are arrayed with symmetry.

- In a crystal, we can designate points whose surrounding are just same to each other. Such points are called the *lattice points*. Lattice points array in same manner as atoms of the crystal. Array of lattice points is called the *lattice*. Each lattice point has an atom or pattern of atoms at same relative position. The atom or pattern of atoms is called the *basis*. So we may say that a crystal is a group of bases occupying the lattice points.

2.1.2 Translational symmetry operation and unit cell

- Symmetry in crystal means that the lattice after a certain operation like rotation or reflection, just overlap perfectly with the initial lattice. The symmetry operations adoptable to natural crystals might be classified in three categories; *translational, point group and space group operations.*

- Translational symmetry operation is to move all the lattice points a certain distance in parallel and may be represented by a vector, T.

$$\vec{T} = n_1\vec{a} + n_2\vec{b} + n_3\vec{c} \tag{2.1}$$

where n's are integers and a, b and c are 'the unit vectors'. The unit vectors are designated in such a way that each of them is one of and smallest of translational symmetry operation by its own and is independent to one another.

- A volume comprised by the unit vectors is called the *unit cell.* You may say that the unit cell can be put together one by one without any vacant space like a brick to form a crystalline solid.

- Crystals in nature are classified into 7 crystal systems by the shape of the unit cell as given in the table below.

Table 2.1 Seven crystal systems

crystal system	lattice parameters
Cubic (立方晶)	a=b=c, $\alpha=\beta=\gamma=90°$
Tetragonal (正方晶)	a=b≠c, $\alpha=\beta=\gamma=90°$
Rhombohedral (三方晶)	a=b=c $\alpha=\beta=\gamma\neq 90°$
Hexagonal (六方晶)	a=b, c $\alpha=\beta=90°$, $\gamma=120°$
Orthorhombic (直方晶)	a≠b≠c≠a $\alpha=\beta=\gamma=90°$
Monoclinic (單斜晶)	a, b, c $\alpha=\beta=90°\neq\gamma$
Triclinic (三斜晶)	a, b, c α, β, γ

2.1.3 Point group symmetry operation

- Unlike the translational symmetry operation, there is another kind of operation during which at least one point stays at the initial position. For instance, rotation about an axis gives no movement on the point at the axis. There are four kinds of such a symmetry operation; rotation, reflection, inverse and rotational inverse.

- *Rotation* is to rotate the lattice around a fixed *axis*. There are 2-fold, 3-fold, 4-fold and 6-fold axes where number of folding means how many times the lattice overlap with the initial lattice during a complete turn of 360°. For instance, a cubic crystal usually has 4-fold, 3-fold and 2-fold axes and it is denoted as 432. A hexagonal crystal has a 6-fold axis as the *principal axis*.

- *Reflection* means that mirror image of the lattice overlaps with the initial lattice. We say that there is a *mirror plane*. Mirror plane is denoted as m. For instance, *4mm* means that the crystal has two mirror planes which are parallel to the principal 4-fold axis. *6/m* means there is a mirror plane perpendicular to the principal 6-fold axis.

- *Inversion* is the symmetry operation about a point, i.e., a centre of symmetry and denoted as $\bar{1}$.

- *Rotational inversion* means a composite operation of

rotation and inverse and denoted as $\overline{2}$, $\overline{3}$, $\overline{4}$ and $\overline{6}$. For instance, $\overline{3}$ means rotation of 120° and inverse give overlapping of lattice. $\overline{2}$ happens to be equivalent to a mirror operation.

- Each operational element by itself and commutation of elements are the members of the point group whose number is 32.

2.1.4 Space group

- Combination of translation and point group symmetry operations gives new symmetry operations, such as gliding plane and screw axis.

- To have a gliding plane means that sliding a half lattice parameter along a plane and then reflection about the plane make the lattice overlap. This operation is denoted as *g*.

- The operation of screw axis is a combination of translation and rotation, For an example, in some materials at every progress of one third of the lattice parameter and rotation of 120°, the lattice overlaps. This case is said to have a screw axis, 3_1. The screw axis 3_2 means that rotation 240°(360×1/3×2) at each translation of one third of the lattice parameter makes the lattice overlap. 2_1, 3_1, 3_2, 4_1, 4_2, 4_3, 6_1, 6_2, 6_3, 6_4 and 6_5 are elements of screw axis. Some of them are equivalent to others.

- Combination between point group elements and gliding planes and screw axes form a new group of symmetry operations, the *space group.* There are 230 kinds of elements in the space group. In a way, we can say that there are 230 different kinds of crystal structures.

2.1.5 Primitive unit cell, centered unit cell and 14 Bravais lattices

- The unit cell defined as a parallelopiped volume comprised by three unit vectors, has lattice points only at the corners. In other words, there is only one lattice point in a unit cell.

- In some cases, the above-defined unit cell does not show

the symmetry of the lattice sufficiently by its intuitive shape. Instead, a pseudo-unit cell which contains lattice point(s) at the center of face(s) or body, represents the symmetry of the lattice much better than the original unit cell. Such *centered unit cells* are widely adopted as unit cell because of convenience though they have more than one lattice points per a cell and offend the original definition of the unit cell. To differentiate, the unit cell with one lattice point per a cell is called the *primitive unit cell.* (Fig 2.1)

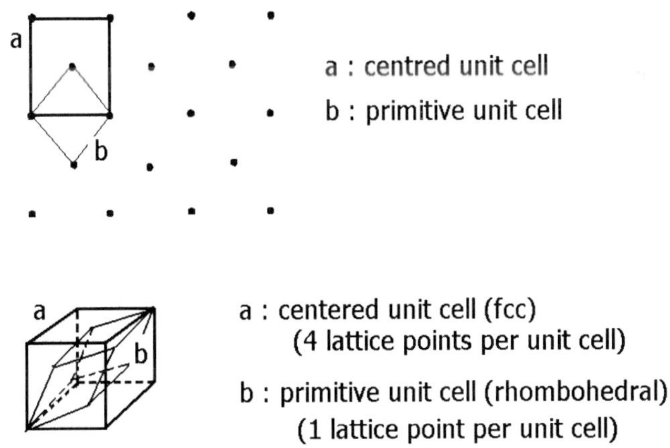

Fig 2.1 Centred unit cell and primitive unit cell.

- Including centered unit cells, seven crystal systems exhibit 14 different kinds of lattices, called *14 Bravais lattices.* They are tabulated below.

Table 2.2 14 Bravais Lattices

Crystal system	Bravais lattice	Remarks
Cubic	P F I	P : simple(primitive)
Tetragonal	P I	F : face centered (all 3 pairs of face)
Rhombohedral	P	I : body centered
Hexagonal	P	C : base centered (only 1 pair of face)
Orthorhombic	P F I C	
Monoclinic	P C	
Triclinic	P	

2.1.6 Notation of position, direction and plane

- The position of a point P is denoted by (x,y,z) where x, y and z are coordinates as fractions of respective unit vectors.(Fig 2.2)

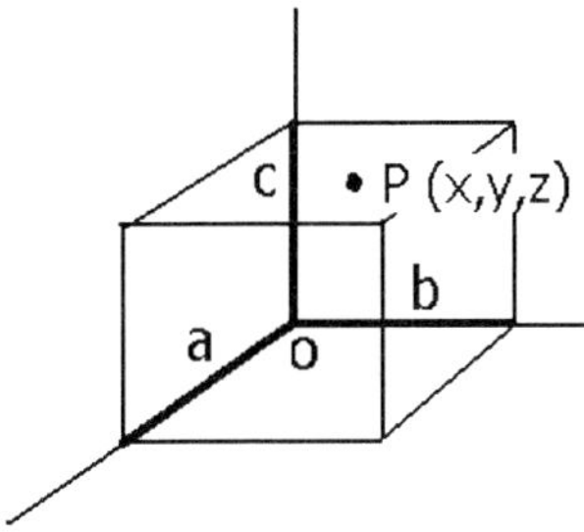

Fig 2.2 Position of a point is expressed as fractions of each coordinates.

- Positions interrelated with addition or subtraction of an multiple of any coordinates are all equivalent position in the crystal. For example,

(1/2,1/2,1/2) ≡ (3/2,1/2,1/2) ≡ (3/2,-1/2,5/2),
(0.0.0) ≡ (1,0,0) ≡ (0,1,1).

- Directions in a crystal are usually denoted by Miller indices. In fact, a direction can be expressed by a vector which begins at the origin and finishes at a point. So a direction can be represented by the position of any point on the arrow line which starts from the origin, [xyz], say. For convenience, we take the indices as simplest integers, [hkl]. For instance, if [xyz] is [2/3,2/3,2/3], then [111] is the [hkl]. h, k and l are called the Miller indices. (Fig 2.3)

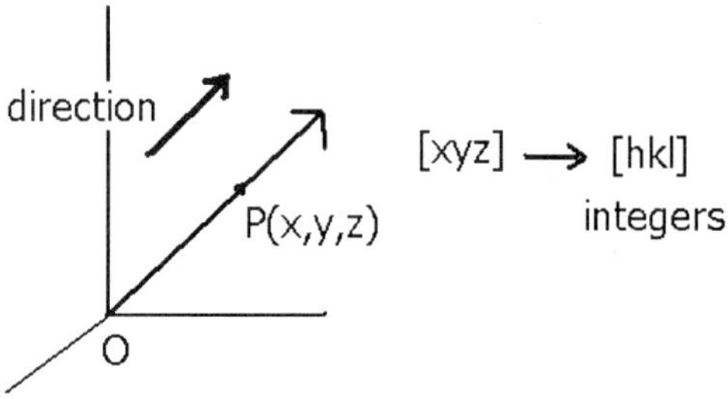

Fig 2.3 Miller indexing for a direction.

- A plane in a crystal is not only one plane, but a group of planes which have same character. The group is characterized by its orientation and interval between neighbouring planes(inter-planar spacing). These two distinguished characters may be determined by the cutting points of three axes by the first member of the plane group after the one passing the origin. (Fig 2.4)

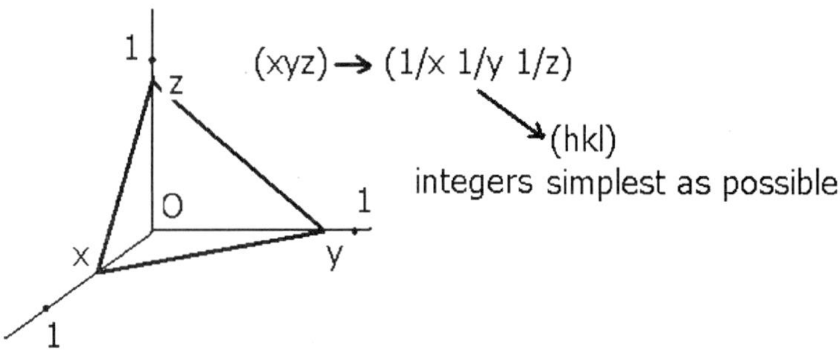

Fig 2.4 Miller indexing for a plane.

- From Fig 2.4, it is seen that (xyz) is surely a method to denote a plane. But x, y, z are not integers in general and sometimes infinities. Miller suggested to express a plane by (1/x 1/y 1/z) instead which gives the Miller indices (hkl). Some examples are;

(xyz)	(1/x 1/y 1/z)	(hkl)
(1/2 1/2 1/2)	(2 2 2)	(222)
(1/2 1/3 1/4)	(2 3 4)	(234)
(11∞)	(1 1 0)	(110)
(111)	(1 1 1)	(111)
(1/2 ∞ ∞)	(2 0 0)	(200)

- We can calculate the *inter-planar spacing*, d, from the Miller indices by a simple geometry. That is,

for a cubic system,

$$d=\frac{a_0}{\sqrt{h^2+k^2+l^2}} \tag{2.2}$$

where a_0 is the lattice parameter, and

for a tetragonal system,

$$d=\frac{1}{\sqrt{\frac{1}{a^2}(h^2+k^2)+\frac{1}{c^2}\cdot l^2}} \tag{2.3}$$

where a and c are also lattice parameters.

- Since a crystal has symmetry of its own. Some directions(or planes) are equivalent to one another. They are said to belong to a *family*. For an example, in a cubic system, directions [100] [010] $[0\bar{1}0]$ $[\bar{1}00]$ [001] and $[00\bar{1}]$ are members of a family. The family name is taken from the representative direction, 100, say, and uses broken brackets, <100>. A family of planes is defined in the same way and denoted by curly brackets, {100}.

- The number of family members is called the multiplicity. The multiplicity depends on the symmetry of the crystal. For a cubic system, 3-fold axis makes permutation of h, k and l to result equivalent direction(plane) and 4-fold axes or mirror planes make change of sign to result in also member of same family. On the other hand, in tetragonal systems, only permutation between h and k and change of signs of h, k and l result in member of same family. For instances, for a cubic crystal, multiplicities of (100) and (211) are 6 and 24, but for a tetragonal crystal, 4 and 16 are the multiplicities, respectively.

- As mentioned above, Miller indices of directions or planes of a family consist of same group of integers, like (110), (101), ($\bar{1}$10)..... . In hexagonal system, however, Miller indices of planes of a family may have different group of indices. For an example, plane (111) has 11 co-family members. They are (2$\bar{1}$1), (1$\bar{2}$1), ($\bar{1}\bar{1}$1), ($\bar{2}$11), ($\bar{1}$21), (2$\bar{1}\bar{1}$), (1$\bar{2}\bar{1}$), ($\bar{1}\bar{1}\bar{1}$), ($\bar{2}$1$\bar{1}$), ($\bar{1}$2$\bar{1}$) and (11$\bar{1}$). (Fig 2.5)

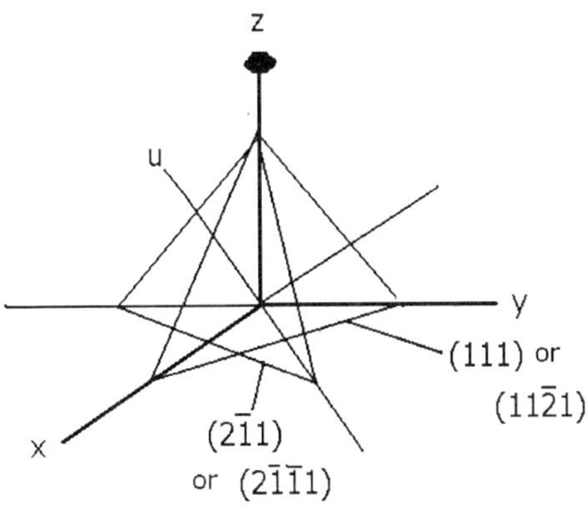

Fig 2.5 Miller indexing and Miller-Bravais indexing for hexagonal system.

- Miller-Bravais indexing for hexagonal system introduces fourth axis, u, which is set as u=-(x+y).(Fig 2.5) Plane or direction is expressed by four indices, as (hkil) where i=-(h+k). Family relation is more easily seen in Miller-Bravais indexing.

2.2 Simple Crystal Structures

(1) Simple cubic

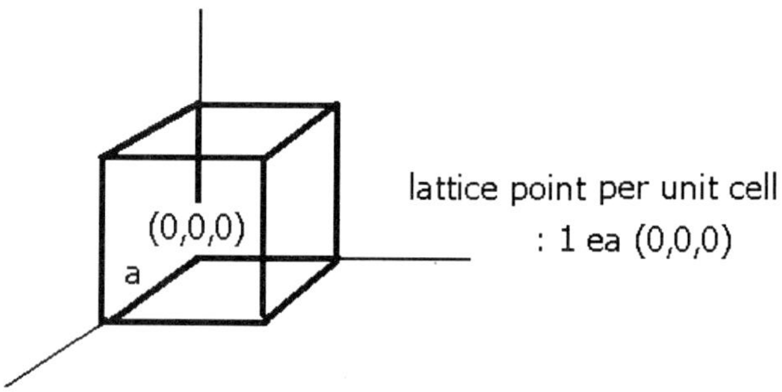

Fig 2.6 Simple cubic structure

- Number of nearest neighbours : 6 (2 at each axis)

  Distance to n.n. : a

- Number of next nearest neighbours :

  12 (1 at each quadrant, 3 planes)

Distance to n.n.n : $\sqrt{2}a$

- Packing ratio(spherical atom and direct contact between n.n. assumed)

$$=\frac{\frac{4\pi}{3}\times(\frac{a}{2})^{3}\times 1}{a^{3}}=\frac{\pi}{6}=0.52$$

(2) Body centred cubic(bcc)

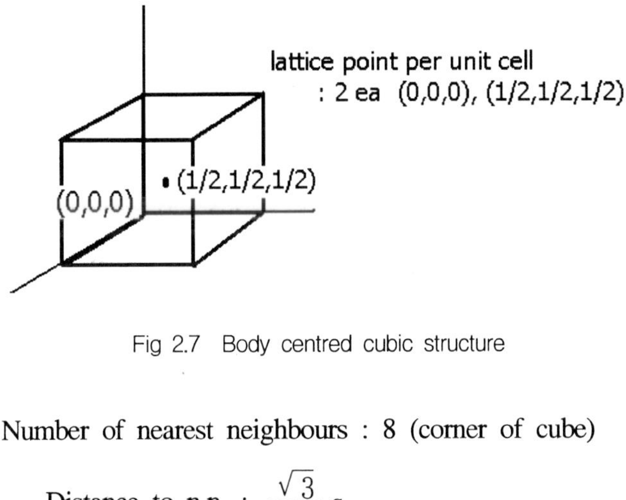

Fig 2.7 Body centred cubic structure

- Number of nearest neighbours : 8 (corner of cube)

  Distance to n.n. : $\frac{\sqrt{3}}{2}a$

- Number of next nearest neighbours : 6 (2 at each axis)

  Distance to n.n.n. : a

- Packing ratio : $\frac{\frac{4\pi}{3}\times(\frac{\sqrt{3}}{4}a)^{3}\times 2}{a^{3}}=\frac{\sqrt{3}\pi}{8}=0.68$

- Close packed plane : (110)

(3) Face centred cubic (fcc)

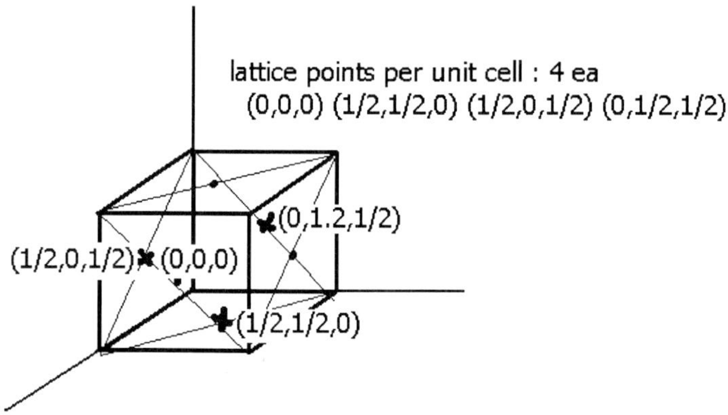

Fig 2.8 Face centred cubic structure

- Number of n.n. : 12 (1 at each quadrant of 3 planes)

  Distance to n.n. : $\frac{\sqrt{2}}{2}a$

- Number of n.n.n. : 6 (2 at each axis)

  Distance to n.n.n. : a

- Packing ratio : $\dfrac{\frac{4\pi}{3}\times(\frac{\sqrt{2}}{4}a)^3\times 4}{a^3}=\frac{\sqrt{2}\pi}{6}=0.74$

- Close packed plane : (111)

(4) NaCl type structure

- NaCl type structure can be interpreted as two inter-penetrating fcc's of which one starts from the origin and second one shifts a half of a unit vector, (1/2,0,0). Different kind of atoms occupy each fcc which is called *sub-lattice*.

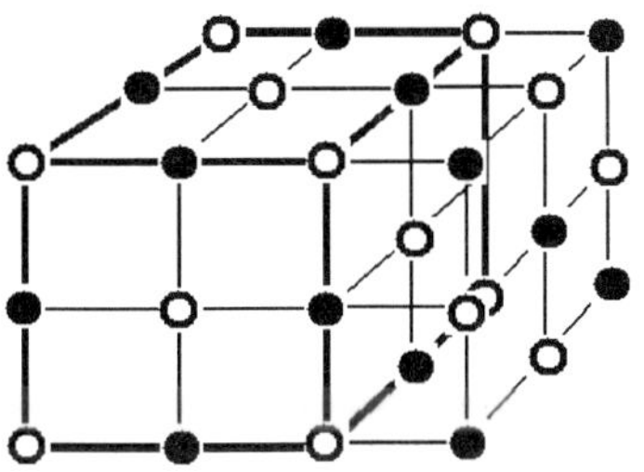

Two interpenetrating fcc's : One starts from (0,0,0)
and second one from (1/2,0,0)

Two sub-lattices and 4 lattice points each per unit cell
A : (0,0,0) (1/2,1/2,0) (1/2,0,1/2) (0,1/2,1/2)
B : (1/2,0,0) (0,1/2,0) (0,0,1/2) (1/2,1/2,1/2)

Fig 2.9 NaCl type structure

- Number of lattice points : Two sub-lattices and four lattice points each.

A (0,0,0) (1/2,1/2,0) (1/2,0,1/2) (0,1/2,1/2)

B (1/2,0,0) (0,1/2,0) (0,0,1/2) (1/2,1/2,1/2)

- Number of n.n. : 6 (different kind of atoms)

  Distance to n.n. : a/2

- Number of n.n.n. : 12 (face center, same kind of atom)

  Distance to n.n.n. : $\frac{\sqrt{2}}{2}a$

(5) CsCl type structure

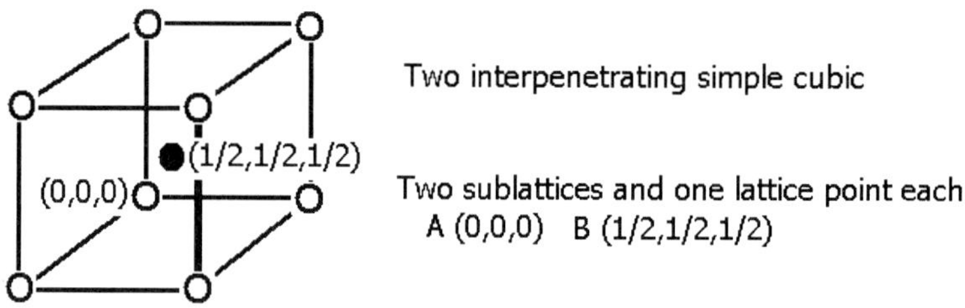

Fig 2.10 CsCl type structure

- CsCl type structure is composed of two inter-penetrating simple cubic lattices; one from the origin and the other from the body centre. This implies that CsCl structure is not a bcc but a simple cubic structure.

- Number of lattice points : Two sub-lattices and one lattice points each.

  A (0,0,0)

B (1/2,1/2,1/2)

- Number of n.n. : 8 (different kind of atom)

  Distance to n.n. : $\frac{\sqrt{3}}{2}a$

- Number of n.n.n. : 6 (same kind of atom)

  Distance to n.n.n. : a

(6) Diamond structure

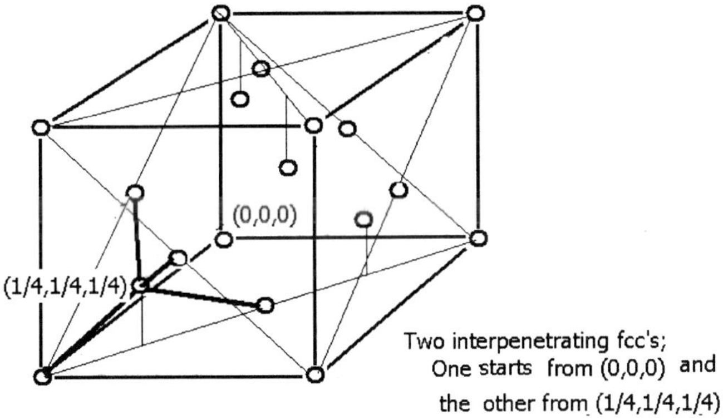

Fig 2.11 Diamond structure

- Diamond structure can be understood as two interpenetrating fcc's. One of them starts at the origin and the other at (1/4,1/4,1/4). There are, therefore, 8 lattice points in a unit cell, all of which are occupied by one kind of atoms.

- Number of lattice points : 8 (4 lattice points per each fcc)

  (0,0,0) (1/2,1/2,0) (1/2,0,1/2) (0,1/2,1/2)

  (1/4,1/4,1/4) (3/4,3/4,1/4) (3/4,1/4,3/4) (1/4,3/4,3/4)

- Number of n.n. : 4 (corner of tetrahedron)

  Distance to n.n. : $\frac{\sqrt{3}}{4}a$

- Number of n.n.n. : 12 (face center)

  Distance to n.n.n. : $\frac{\sqrt{2}}{2}a$

- Packing ratio : $\dfrac{\frac{4\pi}{3}\times(\frac{\sqrt{3}}{8}a)^3\times 8}{a^3}=\frac{\sqrt{3}\pi}{16}=0.34$

(7) Zinc blende(ZnS) type

- ZnS type structure resembles diamond structure, except that two inter-penetrating fcc's are occupied by two kinds of atoms, respectively. That is, there are two sub-lattices with four lattice points each.

  A (0,0,0) (1/2,1/2,0) (1/2,0,1/2) (0,1/2,1/2)

  B (1/4,1/4,1/4) (3/4,3/4,1/4) (3/4,1/4,3/4) (1/4,3/4,3/4)

- Number of n.n. : 4 (different kind of atoms)

Distance to n.n. : $\frac{\sqrt{3}}{4}a$

- Number of n.n.n. : 12 (same kind of atoms)

Distance to n.n.n. : $\frac{\sqrt{2}}{2}a$

(8) Hexagonal close packed (hcp)

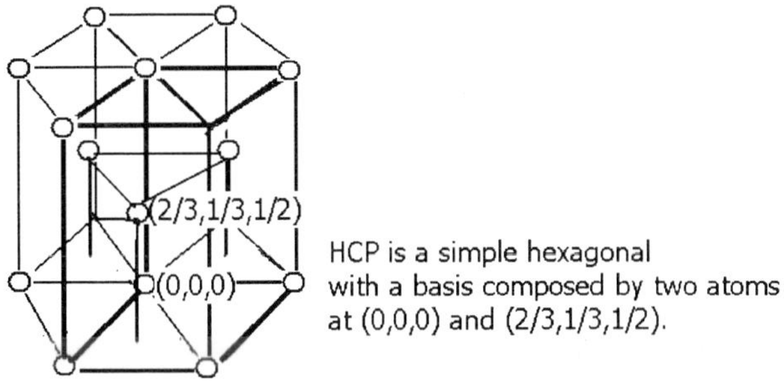

Fig 2.12 HCP structure

- HCP is a simple hexagonal structure and the basis is composed with two atoms at (0,0,0) and (2/3,1/3,1/2).

- In the 'ideal' case, all the nearest neighbours within a basal plane and successive layers are in contact. The c/a ratio is 1.633 for the 'ideal' case. In most cases, however, c/a ratios are larger or less than that of the 'ideal' case.

- Number of n.n. : 6 (co-planar)

  6 (3 each at upper and lower adjacent layer)
- Packing ratio of 'ideal' hcp is same as that of fcc.
- Stacking sequence of hcp is **ABABAB**.... while fcc shows stacking sequence as **ABCABCA**..... .

2.3 Wave and crystal

- In crystals, lots of waves of various kinds exist ; electromagnetic, lattice and electronic waves(matter wave). Waves are moving forward, scattered by atoms, reflected by planes and subject to interference with one another to result in standing waves sometimes. Such phenomena are quite important in understanding properties of solids.

- It will be convenient for dealing with interactions between waves and crystal, if planes and waves are expressed in one space. One of methods to fulfil such purpose is introduction of the *reciprocal lattice.*

2.3.1 Reciprocal lattice

- Unlike points or lines, a number of planes are rather difficult to express in space. Furthermore, a plane denoted by (hkl) means not a single plane but a group of planes.

- A plane (hkl) is characterized by the inter-planar spacing, d, and the orientation which can be defined by the plane normal vector. These two parameter can be represented by a vector whose magnitude is d and direction is that of the plane normal.(Fig 2.13) It also can be represented by a point as the vector is presumed to start from the origin and to finish at the point.

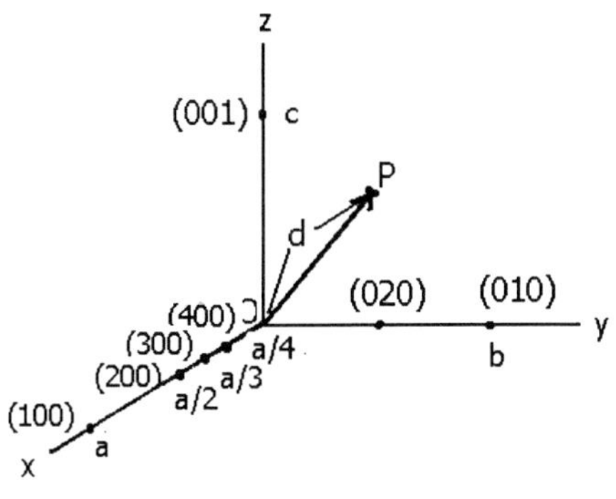

Fig 2.13 A plane can be expressed by a vector or a point.

- The real space expression, however, has difficulties as seen in Fig 2.13, since inter-planar spacings become smaller and smaller and points for planes become densely compact with higher h, k, l. This difficulties can be removed if we introduce the reciprocal space where length is expressed by its inverse value. Directions are expressed in the same way as the real space.

- Now in the reciprocal space, distance is expressed as $2\pi/l$ (the factor, 2π, is introduced to conform with the definition of the wave vector). And a plane is denoted by a vector, G_{hkl}, whose magnitude is $2\pi/d_{hkl}$ and direction is the plane normal, or by a point which is the arrow tip of the vector. The vector is called the *reciprocal lattice vector* and the point the *reciprocal lattice point.* (Fig 2.14)

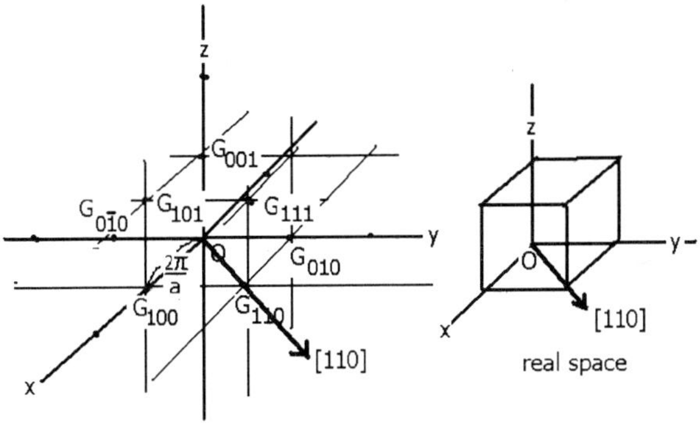

Fig 2.14 Reciprocal lattice for a cubic crystal.
Point Ghkl represents (hkl) plane.

- It is seen in Fig 2.14 that points G_{hkl} form a lattice in the reciprocal space. That is why it is called the reciprocal lattice. It is also seen that the reciprocal lattice has the same symmetry as the real lattice. Directions are expressed in same way in both lattices.

- Waves(electromagnetic, phonon or electron) are characterized by wave length λ and direction of propagation. These two factors can be expressed by the wave vector **k**, which is defined as,

$$k = \frac{2\pi}{\lambda}. \tag{2.4}$$

- Waves, therefore, are represented by k-vectors in the reciprocal lattice without difficulties. In sometimes, waves are denoted by points in the reciprocal lattice as the points mean the arrow tip of the k-vector starting from the origin.

2.3.2 Brillouin zone

- Waves are scattered by particles, atoms or electrons. If there is only one particle, scattered wave propagates into every directions around. But, in practice, there are numerous particles in vicinity and scattered waves from neighbouring particles interfere to make strong waves in particular directions while virtually zero intensity of waves in other directions. Reflection and diffraction are among such phenomena.

- Let's consider one plane of particles where inter-particle distance is smaller than or comparable with the wavelength of incident wave. To make a constructive interference, phase difference between two scattered waves from neighbouring particles should be zero or integral multiple of 2π. This means that path difference between two scattered waves should be zero or integral multiple of the wavelength. With small inter-particle distance, the condition reduces to zero path difference. This condition is expressed as the law of reflection, saying that incident angle and reflection angle should be same.(Fig 2.15)

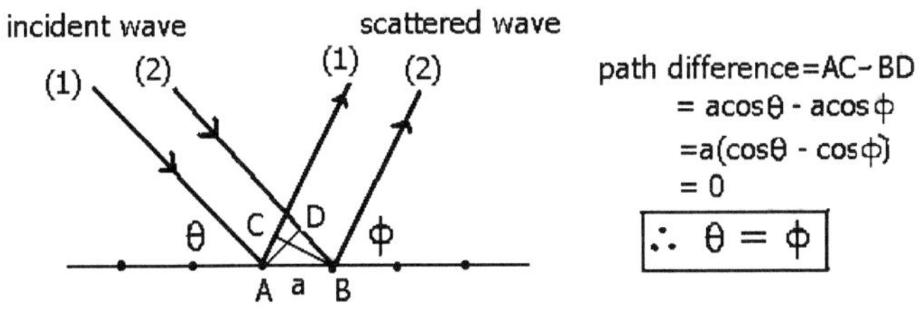

Fig 2.15 Condition for reflection in one plane.

- When a wave reflects from (hkl) plane which is a group of planes, it should satisfy condition for constructive interference between scattered waves from adjacent layers as well as the law of reflection. This is diffraction of waves by crystal plane and the condition is expressed by the Bragg equation.(Fig 2.16)

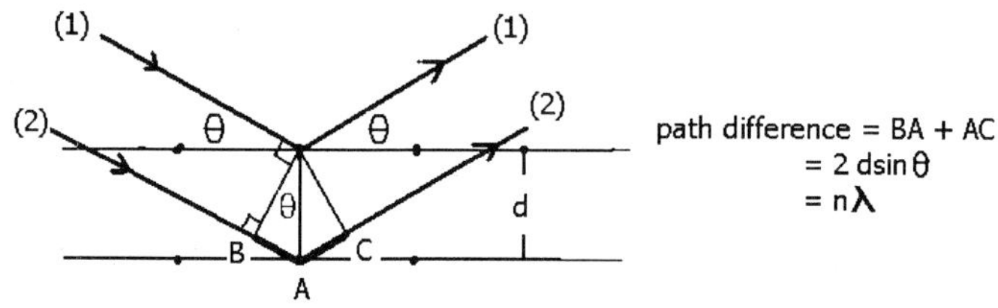

Fig 2.16 Bragg condition for diffraction.

- Bragg equation for diffraction at the plane (hkl) is

$$2d\sin\theta = \lambda. \tag{2.5}$$

Eq(2.5) can be transformed to fit in the reciprocal lattice. That is,

$$\frac{1}{2d\sin\theta} = \frac{1}{\lambda} \rightarrow \frac{2\pi}{2d\sin\theta} = \frac{2\pi}{\lambda} = k$$

$$\frac{2\pi}{d} \times \frac{1}{2} = k \cdot \sin\theta \rightarrow \frac{1}{2} G_{hkl} = k \cdot \sin\theta \tag{2.6}$$

- Eq(2.6) is another expression for the Bragg condition, which can be interpreted as when ksinθ (θ is the angle between the wave and the plane) is just half of the reciprocal lattice vector, G_{hkl}, the wave is reflected by the plane represented by G_{hkl}.(Fig 2.17)

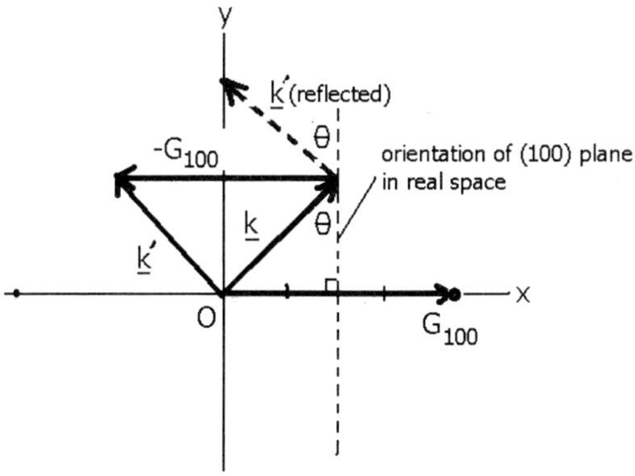

Fig 2.17 Bragg reflection in the reciprocal lattice.

- It is seen that, if the tip of k vector lies on the bisectional plane of G_{hkl}-vector, the wave is reflected into k' vector by the plane (hkl). If the scattering is assumed elastic, from Fig 2.17 we get,

$$|\vec{k}| = |\vec{k'}|$$

$$\therefore \vec{k} - \vec{G} = \vec{k'} \quad . \tag{2.7}$$

Squaring both sides, we obtain,

$$k^2 - 2\vec{k} \cdot \vec{G} + G^2 = k'^2 \quad \text{and} \quad k^2 = k'^2 \quad .$$

$$\therefore \; G^2 - 2\vec{k} \cdot \vec{G} = 0 \; . \tag{2.8}$$

This is another way to express the Bragg condition in the reciprocal space.

- From above discussions, it is concluded that waves are reflected only if the wave vector lies on a bisectional plane of a G vector and otherwise, in other words if tip lies in between bisectional planes, waves pass 'freely' through the crystal without reflections. The zone where 'free-passing' waves lie is called the *Brillouin zone* and Brillouin zones are enclosed by bisectional planes to form polygons. The bisectional planes are called the *Brillouin zone boundary*. So we may say that waves at the Brillouin zone boundary are reflected. Brillouin zones are numbered 1st, 2nd.. with k-value increasing.(Fig 2.18)

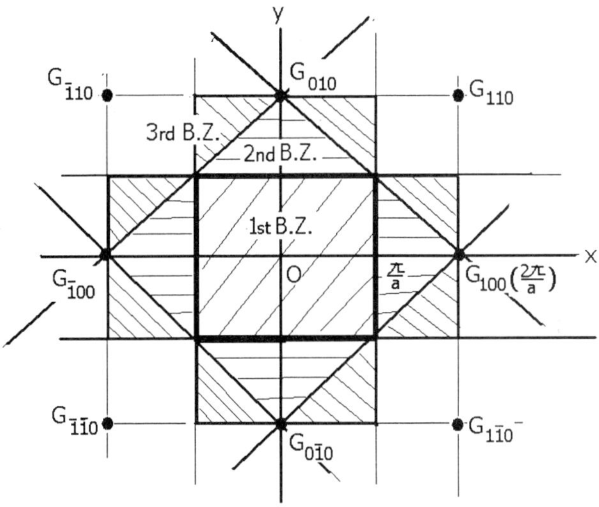

Fig 2.18 Brillouin zones for a two-dimensional square lattice.

Chapter Three

Lattice Wave and Specific heat of Solids

3.1 Harmonic Oscillation of Atoms in Solids

- Atoms in solids are connected to neighbouring atoms by bonding forces and locate at equilibrium positions. When an atom is exerted on by a force, the atom will be displaced from the equilibrium position. If the force is not too strong, the displacement will be within the Hooke's law region. (Fig 1.2) As the force is lifted, the atom will execute an oscillation in a harmonic mode. If the displacement is over the Hooke's law region, the oscillation becomes inharmonic.

- Since atoms are connected to neighbouring atoms, the atomic oscillation will transfer to adjacent atoms successively. In other words, the oscillation travels through the crystal and form a wave. This sort of wave is called the *lattice wave*. For low energy region, like at low temperatures, the lattice waves are well approximated as harmonic waves.

- Atomic oscillations and lattice waves can be excited by various means, such as heat, light, external shock or vibration. Only some waves, however, can sustain through the crystal. These characteristic waves have their own specific modes, sets of frequency and wavelength, which are determined by the mass of atom, interatomic force and lattice parameters.

- We study how these specific modes are determined, for some simple cases - one dimensional uniform solid, monatomic and diatomic lattice. Longitudinal waves (displacement is collinear with the wave direction) are dealt for simplicity.

3.1.1 One dimensional homogeneous line

- When a solid is exerted by a tension, it is elongated. In the Hooke's law region, the tension and elongation show linear proportionality.

$$T(tension) = C \cdot U(displacement) \tag{3.1}$$

where C is called the stiffness of the solid. This macroscopic relation can be expressed in the microscopic relation as,

$$S(stress,\ force/unit\ \ area)$$
$$= Y \cdot \varepsilon(strain,\ displacement/unit\ \ length). \tag{3.2}$$

Here, Y is called the Young's modulus.

- Let's consider a long linear homogeneous solid with density ρ and cross section A.(Fig 3.1) u(x) is the displacement from the equilibrium position at point x. S(x) and ε(x) are stress and strain at point x, respectively.

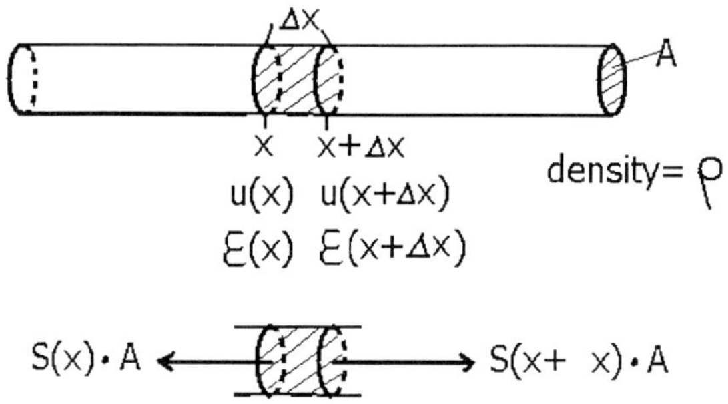

Fig 3.1 Strain and stress in a homogeneous solid.

- Displacement and strain are continuous in the homogeneous solid.

$$U(x+\triangle x)=U(x)+\frac{\partial U}{\partial x}\cdot\triangle x$$

$$\varepsilon(x)=\frac{\partial U}{\partial x}$$

$$\varepsilon(x+\triangle x)=\varepsilon(x)+\frac{\partial \varepsilon}{\partial x}\cdot\triangle x=\frac{\partial U}{\partial x}+\frac{\partial^2 U}{\partial x^2}\cdot\triangle x \qquad (3.3)$$

- Now forces on both ends of the line element, $\triangle x$, are S(x)A and S(x+$\triangle$x)A at left and right hand side, respectively, and the resultant force becomes,

$$F=S(x+\triangle x)\cdot A-S(x)\cdot A$$

$$= Y \cdot [\varepsilon(x+\triangle x)-\varepsilon(x)] \cdot A$$

$$= Y \cdot \frac{\partial^2 U}{\partial x^2} \cdot \triangle x \cdot A \tag{3.4}$$

- We know the Newton's force law,

$$F = ma.$$

Here,

$$m = A \cdot \triangle x \cdot \rho \quad \text{and} \quad a = \frac{\partial^2 U}{\partial t^2}.$$

$$\therefore \; Y \cdot \frac{\partial^2 U}{\partial x^2} \cdot \triangle x \cdot A = A \cdot \triangle x \cdot \rho \cdot \frac{\partial^2 U}{\partial t^2}$$

and we get,

$$\frac{\partial^2 U}{\partial t^2} = \frac{Y}{\rho} \frac{\partial^2 U}{\partial x^2}. \tag{3.5}$$

This is called the wave equation.

- The general solution of the wave equation is,

$$U = U_0 \cdot e^{-i(\omega t - kx)} \tag{3.6}$$

where ω=2πν and k=2π/λ. From Eq (3.5) and (3.6), we obtain,

$$\begin{aligned} -\omega^2 U &= \frac{Y}{\rho} \cdot (-k^2) U \\ \therefore \omega &= \pm\sqrt{\frac{Y}{\rho}} \cdot k = \pm v_0 k \end{aligned} \tag{3.7}$$

- Eq (3.7) gives the relation between ω and k, which is called the *dispersion relation.* The ω-k curve is called the *dispersion curve.* (Fig 3.2) The dispersion relation designates the characteristic lattice waves which are permeable through the solid.

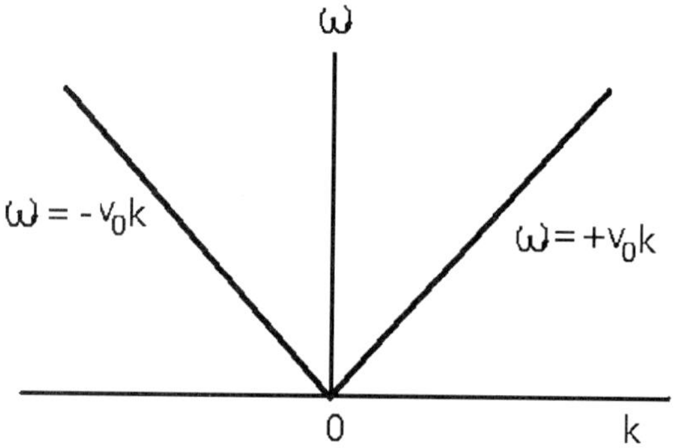

Fig 3.2 Dispersion curve of a homogeneous line.

- From dispersion relation, wave velocity can be obtained. For a single wave, the velocity is the product of frequency and wavelength. that is,

$$v_p = \nu \cdot \lambda = \frac{\omega}{2\pi} \cdot \frac{2\pi}{k} = \frac{\omega}{k} \quad . \tag{3.8}$$

This expression is applicable to most macroscopic systems and often called the phase velocity to differentiate from another concept of velocity, the *group velocity*.

- In microscopic system, we know that the uncertainty principle rules the physical world. It is impossible for a wave to hold a set of ω and k in exact values. Instead, a wave represented by a set of (ω,k) may be interpreted as a interference pattern of a group of numerous waves whose ω- and k-values are near to (ω,k). Naturally, the velocity of the wave (ω,k) should be taken not as the velocity of the individual waves but as the velocity of the 'interference' pattern. This is called the group velocity and is given as,

$$v_g = \frac{d\omega}{dk} \quad . \tag{3.9}$$

- It is seen, from the dispersion relation of Eq (3,7), that $v_g=v_p=v_0$ for a homogeneous solid. In fact, there is no homogeneous solid and masses are concentrated on atoms. But if the wavelength is much larger than the interatomic distance ($\lambda \gg a$), then the wave feels that the solid is homogeneous. In this circumstance the wave can be dealt as macroscopic. When the wavelength is comparable with the interatomic distance, Eq (3.7) is no longer applicable.

3.1.2 One dimensional monatomic lattice

- The simplest model of real solid is the lattice of single kind of atoms. Lattice waves of one dimensional monatomic lattice is studied. Assuming the atomic oscillations are within the Hooke's law region, atoms can be considered to be connected to adjacent atoms by a spring.(Fig 3.3)

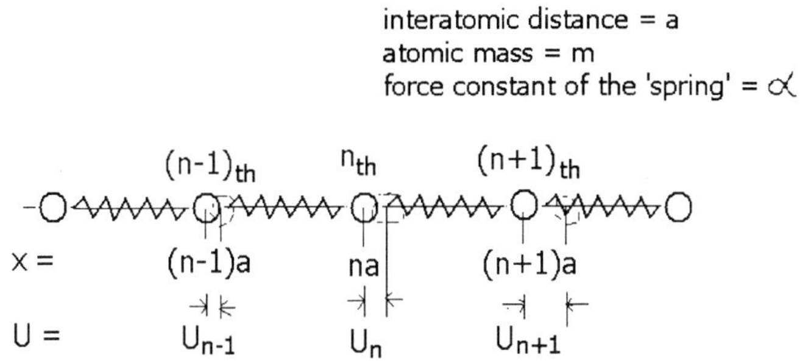

Fig 3.3 One dimensional monatomic lattice.

- Forces on the n-th atom are from neighbouring atoms on right and left hand sides. That is,

$$F=\alpha(U_{n+1}-U_n)-\alpha(U_n-U_{n-1})=\alpha(U_{n+1}-2U_n+U_{n-1})$$

Here, α is the force constant of the 'spring' connecting atoms and related with the Young's modulus as,

$$\alpha\cdot U=Y\cdot\frac{U}{a}\quad\text{or}\quad\alpha=\frac{Y}{a}\ . \tag{3.10}$$

- Using the force law, we get,

$$\alpha(U_{n+1}-2U_n+U_{n-1})=m\frac{\partial^2U_n}{\partial t^2}\ . \tag{3.11}$$

- Now, we assume the motion of atoms is along harmonic waves. That is, displacements of atoms can be written as,

$$U = U_0 \cdot e^{-i(\omega t - kx)} \,. \tag{3.12}$$

- Then,

$$\begin{aligned} U_n &= U_0 \cdot e^{-i(\omega t - kna)} \\ U_{n+1} &= U_0 \cdot e^{-i[\omega t - k(n+1)a]} \\ U_{n-1} &= U_0 \cdot e^{-i[\omega t - k(n-1)a]} \,. \end{aligned} \tag{3.13}$$

- From Eq's (3.11) and (3.13), we get,

$$m \cdot (-\omega^2) \cdot U_0 \cdot e^{-i(\omega t - kna)} = \alpha[e^{ika} - 2 + e^{-ika}] \cdot U_0 \cdot e^{-i(\omega t - kna)}$$

$$\therefore -m\omega^2 = \alpha[2\cos ka - 2] = \alpha[-4\sin^2\frac{ka}{2}]$$

$$\therefore \omega = \pm\sqrt{\frac{4\alpha}{m}} \cdot \sin\frac{ka}{2} = \pm\omega_m \sin\frac{ka}{2} \tag{3.14}$$

This is the dispersion relation for the 1-D monatomic lattice. It is seen that there is a upper limit in the frequency. Fig 3.4 shows the dispersion curve of the one dimensional monatomic lattice.

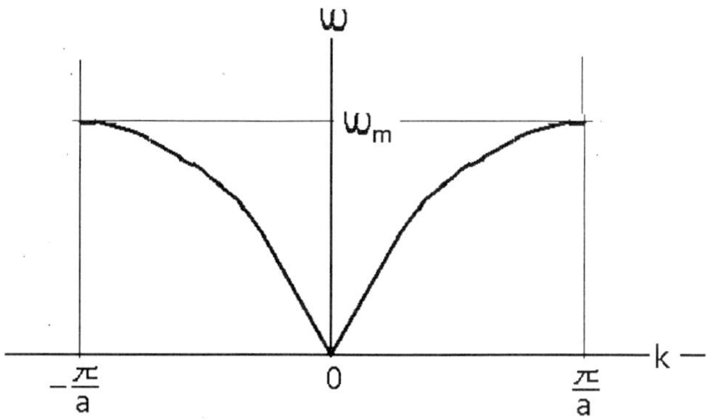

Fig 3.4 Dispersion curve of 1-D monatomic lattice

- The velocity of wave is obtained from Eq 3.14 as a function of k.

$$v_g = \frac{d\omega}{dk} = \pm\sqrt{\frac{4\alpha}{m}} \cdot \frac{a}{2}\cos\frac{ka}{2} = \pm\sqrt{\frac{a \cdot \alpha}{m/a}} \cdot \cos\frac{ka}{2}$$

$$= \pm\sqrt{\frac{Y}{\rho}} \cdot \cos\frac{ka}{2} = \pm v_0 \cos\frac{ka}{2}. \qquad (3.15)$$

- Let's consider two extremes, k→0 and k→±π/a.

(i) When k→0, cos(ka/2)→1.

$$\therefore\ v_g = \pm v_0\ .$$

This is same as that of homogeneous line. It can be easily understood since $k \to 0$ means a wavelength long enough for the wave to feel the interatomic distance nearly zero.

(ii) When $k \to \pm\pi/a$, $\cos(ka/2) \to 0$. Then,

$$v_g = 0 .$$

The velocity is zero. What happens? It is because, at this value of k, the wave satisfies the Bragg condition and reflects from the plane, (100) say. The foregoing and reflected waves interfere to form a standing wave, which is standing, not moving in a way.

- So, the region in k-space between $-\pi/a<k<\pi/a$ corresponds to the first Brillouin zone.

- Waves with $k>\pi/a$ and $k<-\pi/a$ can be reduced to waves with $-\pi/a<k'<\pi/a$ by the relation $k'=k\pm G$. In fact, a wave with $k>\pi/a$ has a wavelength shorter than 2a. In this case, the wave has maxima at empty space which is meaningless. Such waves are well represented by other waves with longer wavelength.(Fig 3.5)

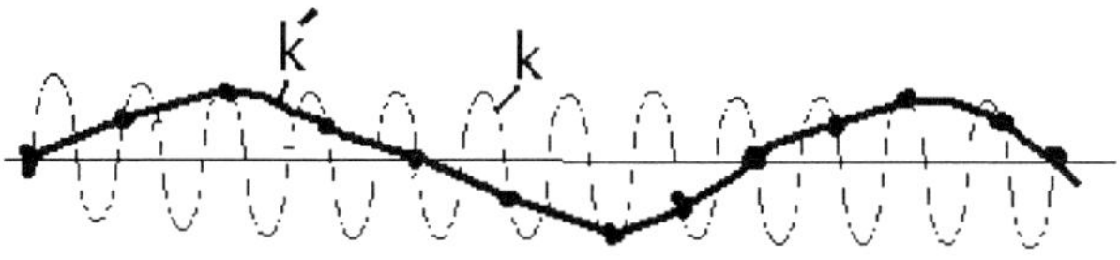

Fig 3.5 Waves with |k|>π/a can be reduced to waves with |k|<π/a.

3.1.3 One-Dimensional Diatomic lattice

- Diatomic lattice holds two atoms in a unit cell. A simple case of the diatomic lattice consists of two kinds of atoms with different masses. Atoms are assumed inter-connected with a spring for harmonic oscillations.(Fig 3.6)

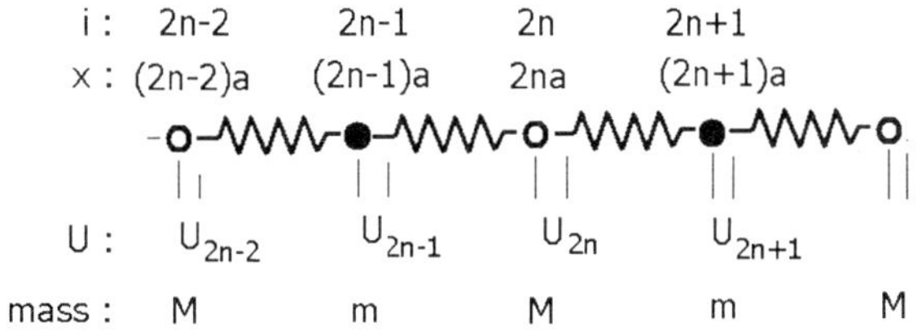

Fig 3.6 One dimensional diatomic lattice.

- Equations of motion for two atomic sites are expressed

as,

$$\alpha[U_{2n+1}-2U_{2n}+U_{2n-1}]=M\frac{\partial^2 U_{2n}}{\partial t^2}$$
$$\alpha[U_{2n}-2U_{2n-1}+U_{2n-2}]=m\frac{\partial^2 U_{2n-1}}{\partial t^2} \qquad (3.16)$$

where α is the force constant of the 'spring' connecting atoms.

- Assuming harmonic oscillations, displacements, U's, for both atomic sites can be expressed in the same form but with different amplitudes. That is,

$$U_{even}=U_{01}\cdot e^{-i(\omega t-kx)}$$
$$U_{odd}=U_{02}\cdot e^{-i(\omega t-kx)} \quad . \qquad (3.17)$$

- Then Eq (3.16) becomes,

$$\alpha[U_{02}\cdot e^{ika}-2U_{01}+U_{02}\cdot e^{-ika}]=-MU_{01}\omega^2$$
$$\alpha[U_{01}\cdot e^{ika}-2U_{02}+U_{01}\cdot e^{-ika}]=-mU_{02}\omega^2 \quad . \qquad (3.18)$$

- And we get a simultaneous equations of U_{01} and U_{02}.

$$\begin{aligned}(M\omega^2 - 2\alpha)U_{01} + 2\alpha\cos ka U_{02} = 0 \\ 2\alpha\cos ka U_{01} + (m\omega^2 - 2\alpha)U_{02} = 0\end{aligned} \quad . \tag{3.19}$$

- The homogeneous simultaneous equations, Eq (3.19), have non-zero solutions for U_{01} and U_{02}, only when

$$\begin{vmatrix}(M\omega^2 - 2\alpha) & 2\alpha\cos ka \\ 2\alpha\cos ka & (m\omega^2 - 2\alpha)\end{vmatrix} = 0 \ . \tag{3.20}$$

- Solving the determinant (Eq 3.20), we get a quadratic equation of ω^2 and the root formula gives us an expression for ω as,

$$\omega^2 = \alpha\left(\frac{1}{M} + \frac{1}{m}\right) \pm \alpha\sqrt{\left(\frac{1}{M} + \frac{1}{m}\right)^2 - \frac{4\sin^2(ka)}{mM}} . \tag{3.21}$$

This equation represents the dispersion relation for a diatomic lattice.

- The dispersion relation can be analyzed for cases of + and - sign at the second term in the right hand side and values of ω can also be evaluated for two extremes, $k \to 0$ and $k \to \pi/2a$(Brillouin zone boundary).

(i) When we take + sign,

$$\omega = [2\alpha(\frac{1}{M} + \frac{1}{m})]^{\frac{1}{2}} \quad \text{at } k \to 0 , \qquad (3.22)$$

and

$$\omega = [2\alpha \cdot \frac{1}{m}]^{\frac{1}{2}} \quad \text{at } k \to \pi/2a . \qquad (3.33)$$

(ii) When we take - sign,

$$\omega = \pm\sqrt{\frac{2a^2\alpha}{M+m}} \cdot k = \pm v_0 k \quad \text{at } k \to 0, \qquad (3.34)$$

where $\sqrt{\frac{2a^2\alpha}{M+m}} = \sqrt{\frac{a\alpha}{\frac{M+m}{2a}}} = \sqrt{\frac{Y}{\rho}} = v_0$.

And

$$\omega = [2\alpha \cdot \frac{1}{M}]^{\frac{1}{2}} \quad \text{at } k \to \pi/2a . \qquad (3.35)$$

- Each case shows its own dispersion curve and we get two separate dispersion curves, which are usually called 'branches'.(Fig 3.7)

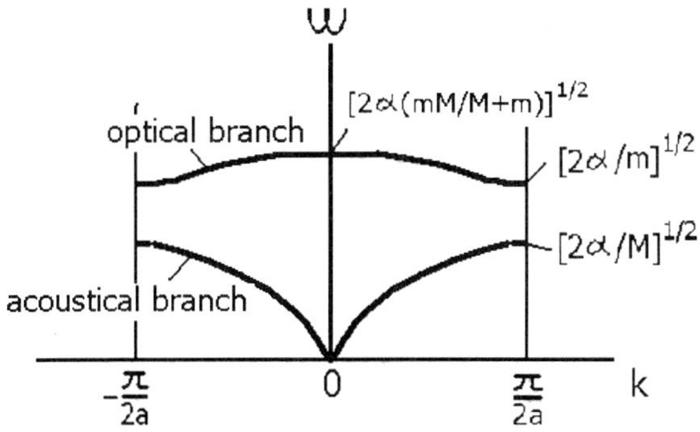

Fig 3.7 Dispersion curves of a diatomic lattice. Optical and acoustical branches.

- In waves of optical branch, m and M oscillate in opposite direction to each other, which can be imagined in ionic crystals excited by electromagnetic waves. On the other hand, for acoustical branch, m and M oscillate in same direction, which can be envisaged in crystals excited by a pressure wave like sound wave.

3.2 Quantization of Lattice Waves

- Lattice waves are confined within a solids which is not infinite in size. This spatial limitation of lattice waves implies the quantization of waves. Quantization of waves which appears as discrete values of wave number, frequency and so energy, can be induced mathematically by the boundary conditions.

- There are two types of boundary conditions for the lattice waves: the fixed boundary condition and the periodic boundary condition. The fixed boundary condition starts from the assumption that the surfaces(x=0 and x=L for a solid of length L) are fixed and not moving. That is,

$$U(0)=U(L)=0. \tag{3.36}$$

Waves have standing character rather than travelling mode in the fixed boundary condition. On the other hand, the periodic boundary condition imposes a symmetrical nature to the waves such as,

$$U(x+L)=U(x)\ . \tag{3.37}$$

- Adopting the periodic boundary condition, we get

$$\begin{aligned} U(x+L) &= U_0 \cdot e^{-i(\omega t - kx - kL)} \\ &= U(x) = U_0 \cdot e^{-i(\omega t - kx)} \end{aligned}$$

and a simple result as,

$$e^{ikL}=1 \ . \tag{3.38}$$

$$\therefore \ kL=2n\pi \quad or \quad k=\frac{2\pi}{L}\cdot n \ , \quad n=\pm1, \pm2, \cdots \tag{3.39}$$

- It is seen that k is quantized and then ω is also quantized through the dispersion relation. E is also quantized since E= ℏω.

- So, waves are now expressed as,

$$U_n = U_0 \cdot e^{-i(\omega_n t - k_n x)} \tag{3.40}$$

which is characterized by a set (ω_n, k_n). We call this a mode.

- Number of modes in a solid can easily calculated(see Fig 3.8). That is,

$$N=\frac{\frac{\pi}{a}}{\frac{2\pi}{L}}\times 2=\frac{L}{a}=Nmber \quad of \quad atoms, \ N \tag{3.41}$$

- Actual total number of modes is 3N instead of N, since there are two transverse waves in addition to the longitudinal wave to make three kinds of waves in total.

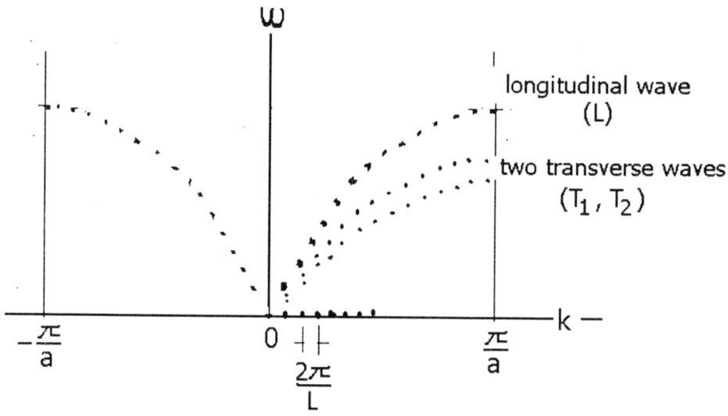

Fig 3.8 Quantization of lattice waves and 3N modes.

- So far one dimensional lattice has been dealt with. Same line of approach, however, is also applicable to a three dimensional lattice as well. k-vector is quantized in 3D crystal as,

$$\vec{k} = k_x \vec{i} + k_y \vec{j} + k_z \vec{k}$$

$$k_x = \frac{2\pi}{L} \cdot n_x,\ k_y = \frac{2\pi}{L} \cdot n_y,\ k_z = \frac{2\pi}{L} \cdot n_z$$

$$(n_x, n_y, n_z = 0, \pm 1, \pm 2, \cdots) \tag{3.41}$$

Here cubic lattice is assumed for simplicity.

- One k-value has a volume of $(\frac{2\pi}{L})^3$ in k-space and the volume of a Brillouin zone in k-space is $(2 \cdot \frac{\pi}{a})^3$. So the total number of k-value's is

$$\frac{(2 \cdot \frac{\pi}{a})^3}{(\frac{2\pi}{L})^3} = \frac{L^3}{a^3} = N$$: Number of atoms in the crystal.

∴ Total number of modes = 3N : one longitudinal and two transverse waves.

3.3 Specific Heat of Solid

3.3.1 Dulong-Petit's law

- The specific heat of a solid is defined as,

$$C = \frac{\triangle Q}{\triangle T} \quad \text{and} \quad \triangle Q = \triangle U + P \triangle V.$$

If the volume remains constant, i.e., $\triangle V=0$, then the constant-volume specific heat becomes,

$$C_v = \left(\frac{\partial Q}{\partial T}\right)_v = \frac{\partial U}{\partial T} \quad (3.42)$$

Here, U is the internal energy of the solid.

- The internal energy of a solids is the sum of kinetic and potential energies of particles in the solids, such as atoms, molecules, free electrons. Among them, the main part is due to atomic vibrations in the form of lattice waves.

- Dulong-Petit's empirical law says that C_v becomes virtually constant as 3R/mole at high temperatures for all solids. In practice, C_v decreases as temperature decreases to low temperatures and becomes proportional to T^3 as T approaches to absolute zero.(Fig 3.9)

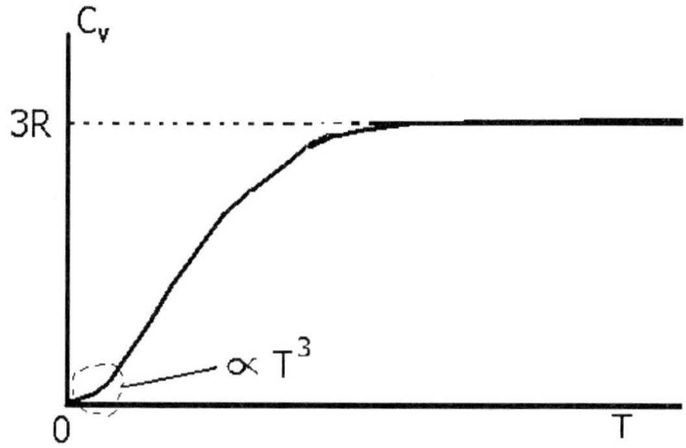

Fig 3.9 Typical Cv-T curve of a solid.

3.3.2 Theory of specific heat of solids

- To understand the specific heat, we have to know the internal energy as a function of temperature. It is helpful to approach the case of real solids through several steps.

(1) Ideal gas model

- Solids are assumed to consist of 'free' point atoms of one kind. This picture is too far away from the real solid. The procedures to get the specific heat in this model is, however, applicable to the case of real solids.

- In this model, atoms have kinetic energies only since they are 'free'. That is,

$$E=\frac{1}{2}mv^2=\frac{1}{2m}p^2=\frac{1}{2m}(p_x^2+p_y^2+p_z^2)$$

and

$$U=\sum_{i=1}^{N}E_i=N\cdot\overline{E_i}$$

$$=N\cdot\frac{\int\int\int\frac{1}{2m}(p_x^2+p_y^2+p_z^2)\cdot e^{-\frac{E}{k_BT}}dp_xdp_ydp_z}{\int\int\int e^{-\frac{E}{k_BT}}dp_xdp_ydp_z}. \quad (3.43)$$

$$=N\cdot\frac{3}{2}k_BT=\frac{3}{2}RT \quad (per\ \ mole: N_A\cdot k_B=R)$$

Here, average is obtained by the Boltzmann distribution.

$$\therefore\ C_v=\frac{\partial U}{\partial T}=\frac{3}{2}R\ (per\ \ mole). \quad (3.44)$$

- Such a discrepancy with the real case is expected right from the beginning.

(2) Classical harmonic oscillators

- In this model, atoms are considered as harmonic oscillator about the equilibrium positions. The oscillators are assumed to be free from others. now each oscillator has kinetic and potential energies. That is,

$$E_i = \frac{1}{2m}(p_x^2 + p_y^2 + p_z^2) + \frac{1}{2}\alpha(u_x^2 + u_y^2 + u_z^2), \qquad (3.45)$$

where α is the force constant between atoms and u's are displacements in x, y and z directions.

- Then the mean value of the energy becomes,

$$\overline{E} = \frac{\int\int\int\int\int\int [\frac{1}{2m}(p_x^2 + p_y^2 + p_z^2) + \frac{\alpha}{2}(u_x^2 + u_y^2 + u_z^2)] e^{-\frac{E}{k_BT}} dp_x dp_y dp_z du_x du_y du_z}{\int\int\int\int\int\int e^{-\frac{E}{k_BT}} dp_x dp_y dp_z du_x du_y du_z}$$

$$= 3k_B T \quad (3.46)$$

$$\therefore\ U = N_A \cdot \overline{E} = 3N_A k_B T = 3RT$$

$$\therefore\ C_v = \frac{\partial U}{\partial T} = 3R \quad (per\ \ mole)\ . \qquad (3.47)$$

- We see that Eq (3.47) is the Dulong-Petit's law. This model, however, cannot explain the decrease of specific heat at low temperatures. The discrepancy is mainly due to the assumption that all oscillation is possible at lowest temperatures no matter what the frequency is. In other words, the oscillations hold continuous energies. But things are not like that. We know that lattice waves are quantized in modes and energies.

(3) Einstein model of quantized harmonic oscillator

- Einstein proposed two assumptions for the specific heat theory of solids; the first one is that energy of a lattice wave should be an integral multiple of $\hbar\omega$, which seems reasonably adoptable.

$$E = n\hbar\omega. \ (n=0,\ 1,\ 2,\ .....) \tag{3.48}$$

- The second assumption, which is apparently unreasonable but helpful for mathematical treatment, says that all the lattice waves have the same frequency, ω_E, called the Einstein frequency.

- Mean value of energy of a lattice wave is obtained as,

$$\overline{E}=\frac{\sum_{n=1}^{\infty} n\hbar\omega_E \cdot e^{-\frac{n\hbar\omega_E}{k_BT}}}{\sum_{n=1}^{\infty} e^{-\frac{\hbar\omega_E}{k_BT}}}=\frac{1}{e^{\frac{\hbar\omega_E}{k_BT}}-1}\cdot\hbar\omega_E$$

$$=\overline{n}\hbar\omega_E \qquad (3.49)$$

Then the internal energy becomes

$$U=3N_A\cdot\overline{E}=3N_A\cdot\overline{n}\hbar\omega_E=3N_A\frac{\hbar\omega_E}{e^{\frac{\hbar\omega_E}{k_BT}}-1}, \qquad (3.50)$$

where total number of modes per mole is $3N_A$.

- Eq (3.50) reduces to simpler forms at two extremes; high and low temperatures.

(i) At high temperatures where $k_BT >> \hbar\omega_E$,

$$e^{\frac{\hbar\omega_E}{k_BT}}\cong 1+\frac{\hbar\omega_E}{k_BT} \quad .$$

Then,

$$U = 3N_A k_B T = 3RT \quad .$$

$$\therefore \ C_v = \frac{\partial U}{\partial T} = 3R \quad .$$

This seems explaining the Dulong-Petit rule well.

(ii) At low temperatures where $k_B T << \hbar\omega_E$, we can neglect (-1) in the denominator of Eq(3.50) and we get,

$$U = 3N_A \cdot \hbar\omega_E \cdot e^{-\frac{\hbar\omega_E}{k_B T}} .$$

$$\therefore \ C_v = 3N_A \cdot \hbar\omega_E \cdot \frac{\hbar\omega_E}{k_B T^2} \cdot e^{-\frac{\hbar\omega_E}{k_B T}}$$

$$= 3R \cdot \left(\frac{\hbar\omega_E}{k_B T}\right)^2 \cdot e^{-\frac{\hbar\omega_E}{k_B T}} \qquad (3.51)$$

Eq (3.51) shows decrease of C_v at low temperatures but much steeper than real measurements, which is expected from the wrong assumption of fixed frequency for lattice waves.

(4) Debye theory of specific heat of solids

- Debye suggested two assumptions which is a bit closer to real crystal.

  (i) Energies of lattice waves are quantized as $E=n\hbar\omega$.
  (ii) Frequencies of lattice waves form a spectrum rather than a fixed frequency. The dispersion relation is assumed as that of a homogeneous solid. That is, $\omega = v_0 k$. This assumption again helps mathematical treatment but differs from the real apparently.

- Now the total internal energy U is given by,

$$U = \sum_{i=1}^{3N} E_i \tag{3.52}$$

where E_i is the energy of i-th mode of wave.

- Eq (3.52) can be rewritten as,

$$U = \sum_{total\ mode}^{3N} g(\omega) \cdot E(\omega). \tag{3.53}$$

Here, g(ω) is the number of modes of which frequencies are in between ω and ω+△ω and E(ω) is the mean energy of these waves.

- For the limit of △ω→0, g(ω)dω represents the number of modes whose frequencies are in between ω and ω+dω. g(ω) is often called the density of modes. Then Eq (3.53) becomes an integral form instead of summation.

$$U=\int_0^{\omega_D} g(\omega)E(\omega)d\omega \tag{3.54}$$

where ω_D is the frequency up to which total number of modes becomes 3N.

- The density of modes, g(ω), is obtained straight forwardly for a homogeneous solid. From Fig 3.10, we can get an expression for g(k)dk for a cubic body with edge L as,

$$g(k)dk=\frac{4\pi k^2 dk}{(\frac{2\pi}{L})^3}\times 3. \tag{3.55}$$

Here, the multiplication factor 3 is for three modes per a

k-value(one longitudinal and two transverse).

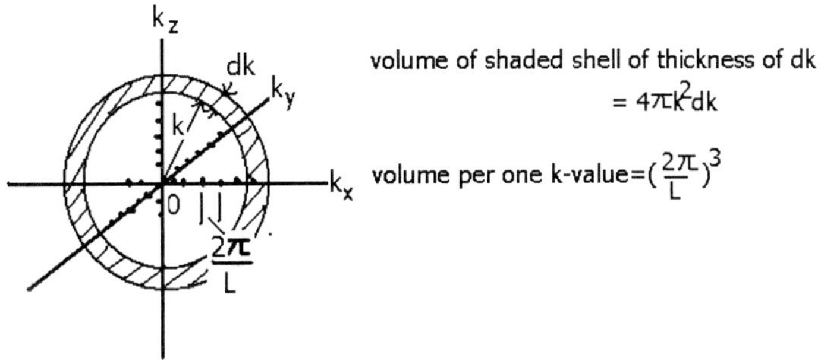

Fig 3.10 g(k)dk is number of k-value in the shaded shell multiplied by 3.

- From the assumed dispersion relation, we get,

$$\omega = v_0 k \qquad \text{and} \qquad d\omega = v_0 dk \ ,$$

$$g(k)dk = \frac{4\pi\left(\frac{\omega}{v_0}\right)^2 \frac{d\omega}{v_0}}{\left(\frac{8\pi^3}{V}\right)} \times 3 = \frac{3V\omega^2 d\omega}{2\pi^2 v_0^3} . \tag{3.56}$$

$$\therefore \ U = \frac{3V}{2\pi^2 v_0^3} \int_0^{\omega_D} \frac{\hbar\omega^3 d\omega}{e^{\frac{\hbar\omega}{k_B T}} - 1} \tag{3.57}$$

- Here, ω_D can be obtained from the following relations.

$$\int_0^{\omega_D} g(\omega)d\omega = 3N.$$

Then we get,

$$\omega_D = v_0(6\pi^2 N_0)^{\frac{1}{3}} \tag{3.58}$$

and

$$k_D = \frac{\omega_D}{v_0} = (6\pi^2 N_0)^{\frac{1}{3}} \tag{3.59}$$

where N_0 = N/V (number of atoms per unit volume).

- The Debye temperature, Θ_D, is defined as

$$k_B \Theta_D = \hbar \omega_D$$

or

$$\Theta_D = \frac{\hbar v_0}{k_B}(6\pi^2 N_0)^{\frac{1}{3}} \tag{3.60}$$

or roughly,

$$\Theta_D \propto \sqrt{\frac{Y}{\rho}} \cdot N_0^{\frac{1}{3}} \text{ again more roughly } \propto \sqrt{\frac{Y}{M}} . \quad (3.61)$$

- At high temperatures,

$$e^{\frac{\hbar\omega}{k_B T}} \cong 1 + \frac{\hbar\omega}{k_B T}$$

and Eq 3.57 becomes

$$U = 3VN_0 k_B T = 3N_A k_B T(per\ mole) = 3RT \quad (3.62)$$

and we see that

$$C_v = 3R \ \ (per\ \ mole).$$

- At low temperatures, probabilities for excitations of lattice waves of higher frequencies are practically so negligible that we can replace the upper limit of the integral of Eq (3.57) with the infinity(∞). Then we can use the integral formula, the Riemann zeta function, as

$$\int_0^{\infty} \frac{x^{s-1}dx}{e^x - 1} = (s-1)! \sum_{n=1}^{\infty} \frac{1}{n^s}$$

and for s=3,

$$\int_0^\infty \frac{x^3 dx}{e^x - 1} = \frac{\pi^4}{15} \quad .$$

Then Eq (3.57) becomes,

$$U = \frac{3V}{2\pi^2} \cdot \frac{(k_B T)^4}{\hbar^3 v_0^3} \cdot \frac{\pi^4}{15} = \frac{3}{5}\pi^4 k_B N T\left(\frac{T}{\Theta_D}\right)^3. \qquad (3.63)$$

and

$$C_v = \frac{\partial U}{\partial T} = \frac{12}{5}\pi^4 k_B N\left(\frac{T}{\Theta_D}\right)^3 = \frac{234R}{\Theta_D^3} \cdot T^3 \ per\ mole. \qquad (3.64)$$

- Debye theory explains the Dulong-Petit rule at high temperatures and T^3 dependence at low temperatures. C_v for real crystals, however, should be obtained by applying the real dispersion relation instead of the homogeneous line's relation. Numerical methods are employed for solutions.

3.3.3 Phonon and zero point energy

- The quantum mechanics shows that the energy of a lattice wave is quantized as,

$$E=\left(n+\frac{1}{2}\right)\hbar\omega, \quad n=0,1,2,\cdots \quad . \tag{3.65}$$

- Lattice wave is sort of energy transport. Eq 3.65 indicates that the unit of energy transported in the form of lattice wave is $\hbar\omega$. The energy particle of the lattice wave is called the *phonon*. When temperature increases, lattice waves become more active and this can be expressed as number of phonons increasing. In fact, at high temperature mean number of phonons, $\overline{n}$, increases linearly with temperature. That is, from Eq 3.49, it is seen that

$$\overline{n}=\frac{1}{e^{\frac{\hbar\omega}{k_BT}}-1}\cong\frac{k_BT}{\hbar\omega}, \quad when\ \ k_BT\gg\hbar\omega. \tag{3.66}$$

- It can be shown that total number of phonons per unit volume is given by,

$$N_{phonon} = 3N(\frac{3}{2} \cdot \frac{T}{\Theta_D}) \tag{3.67}$$

as long as $T >> \Theta_D$, where N is number of atoms per unit volume.

- As temperature goes to absolute zero, number of phonons, $\overline{n}$, becomes zero. Each mode of lattice wave still holds energy, $E_0 = 1/2\hbar\omega$. In other words, atoms continues vibrating even at T=0 K. E_0 is called the zero point energy. Such zero point oscillations are believed helping the appearance of superconductivity at very low temperatures near absolute zero.

3.3.4 Thermal conductivity

- Internal energy diffuses from regions of higher temperature meaning higher energy concentration to regions of lower temperature. Heat conduction per unit area and unit time (energy current density) can be expressed by the following relation,

$$Q = -\chi \frac{dT}{dx} \tag{3.68}$$

where κ is a material constant, called the *thermal conductivity.*

- Energy is transported by energy carrying particles, electron and phonon, say. For insulators, phonon is the major carrier. A rough and qualitative derivation of thermal conductivity of an insulator based on a simple kinetic model is given here. Let phonon be the ideal gas with density of N. The mean velocity and mean free path of phonon are v and λ, respectively. The average energy of a phonon is taken as $3/2k_BT$. Then heat flow through a unit cross section can be expressed in a simple form. (Fig 3.11)

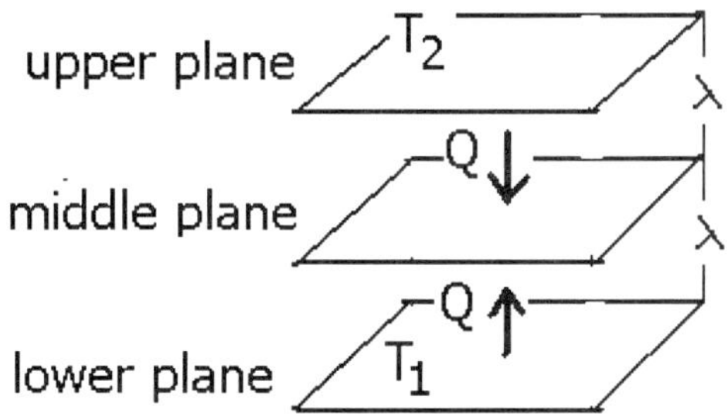

Fig 3.11 Schematic illustration for the kinetic theory of thermal conductivity. Heat flow on the middle plane is analysed.

- Since the inter-planar distance is the mean free path, phonons which leave upper or lower plane will reach the middle plane. The current density of phonon, both downward and upward, is $\frac{1}{6}(N \cdot v)$. (one of six directions) Then energy current density, up and down, can be wriiten as,

$$Q_{down} = \frac{1}{6}(N \cdot v)(\frac{3}{2}k_B T_2)$$
$$Q_{up} = \frac{1}{6}(N \cdot v)(\frac{3}{2}k_B T_1) \qquad (3.69)$$

and the net flow of heat is,

$$Q = Q_{down} - Q_{up} = \frac{1}{4}Nvk_B(T_2 - T_1). \qquad (3.70)$$

downward, assuming $T_2 > T_1$.

- It is seen from Fig 3.11 that the temperature gradient is

$$\frac{dT}{dx} = \frac{T_2 - T_1}{2\lambda} \quad .$$

Then Eq 3.70 can rewritten as,

$$Q = -\frac{1}{2} Nvk_B\lambda \frac{dT}{dx}$$

or

$$\chi = \frac{1}{2} Nk_B v\lambda = \frac{1}{3} C_v v\lambda \qquad (3.71)$$

where C_v is taken as the specific heat of the ideal gas per unit volume, $3/2Nk_B$.

- Although the above approach is too rough to explain the conductivity of real crystals, it gives a good qualitative picture for the thermal conductivity of solids. Thermal conductivities at two extremes, $T \gg \Theta_D$ and $T \ll \Theta_D$, are well explained by the above simple theory.

- When $T \gg \Theta_D$, we can say that

$$\lambda \propto \frac{1}{N_p},$$

where N_p is the density of phonon, and

$$N_p \propto \left(\frac{T}{\Theta_D}\right)$$

while C_v remains constant as $3Nk_B$. As a result, we get

$$\chi = constant \cdot \left(\frac{\Theta_D}{T}\right) . \tag{3.72}$$

This is well in agreement with experiments. (Fig 3.12)

- In the low temperature limit, we know that

$$N_p \propto \left(\frac{T}{\Theta_D}\right)^3$$

and

$$C_v \propto \left(\frac{T}{\Theta_D}\right)^3 .$$

- These relations imply that κ would be constant at low temperatures. In practice, however, κ increases as T goes down. This is because at very low temperatures the anharmonicity of phonon decreases and the mean free path increases more than expected by $(1/N_p)$ relation. λ increases and eventually equals to the specimen dimensions and κ shows saturation. If temperature goes down further, κ decreases since C_v approaches zero.

- For non-crystalline materials like glass, the mean free path is determined by the atomic disorder not by phonon-phonon scattering and usually limited to the average interatomic distance. So we may expect κ varies with temperature in the same way as C_v does. Fig 3.12 shows some examples.

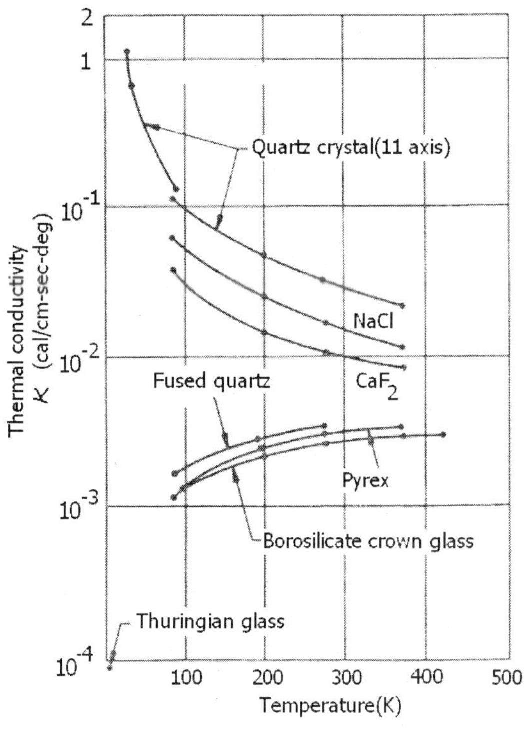

Fig 3.12 Temperature dependence of thermal conductivity.

3.3.5 Thermal expansion

- It is well known that solids expand as temperature goes up. Expansion means that the average inter-atomic distance increases. We know that harmonic oscillation is symmetrical about the equilibrium position and does not incur expansion. Thus to understand the thermal expansion, inharmonic component of lattice vibration should be analysed.

- In order to account for thermal expansion, we introduce a potential function with asymmetric terms which is closer approximation to the real potential curve. Taking the equilibrium position as the origin,

$$U(x) = ax^2 - bx^3 - cx^4 \tag{3.73}$$

where x is a displacement from the equilibrium position representing thermal expansion.

- The average value of x at a given temperature is calculated using the Boltzmann distribution function.

$$\langle x\rangle = \frac{\int_{-\infty}^{\infty} x\cdot \exp[\frac{-(ax^2-bx^3-cx^4)}{k_BT}]dx}{\int_{-\infty}^{\infty} \exp[\frac{-(ax^2-bx^3-cx^4)}{k_BT}]dx}. \tag{3.74}$$

Assuming small inharmonic terms, we can use the following approximation,

$$\begin{aligned}\exp[\frac{-(ax^2-bx^3-cx^4)}{k_BT}] &= \exp[\frac{-ax^2}{k_BT}]\cdot \exp[\frac{bx^3+cx^4}{k_BT}] \\ &\simeq \exp[\frac{-ax^2}{k_BT}](1+\frac{bx^3}{k_BT}+\frac{cx^4}{k_BT})\end{aligned}.$$

Then Eq 3.74 becomes,

$$\langle x\rangle = \frac{\int_{-\infty}^{\infty} \exp[\frac{-ax^2}{k_BT}](x+\frac{bx^4}{k_BT}+\frac{cx^5}{k_BT})dx}{\int_{-\infty}^{\infty} \exp[\frac{-ax^2}{k_BT}]dx} \tag{3.75}$$

where inharmonic terms are neglected at the denominator.

- Eq 3.75 gives,

$$\langle x\rangle = \frac{3}{4}\cdot\frac{b}{a^2}k_BT. \tag{3.76}$$

The expansion coefficient is given by

$$\alpha = \frac{\partial x}{\partial T} = \frac{3k_B b}{4a^2} \tag{3.77}$$

which is temperature-independent. This is not applicable at low temperatures where quantum nature of lattice vibration should be taken into account.

- Eq 3.76 is rewritten as

$$\langle x \rangle = \frac{3}{4} \cdot \frac{b}{a^2} \overline{E}. \tag{3.78}$$

If we adopt the simple (Einstein) model,

$$\overline{E} = \frac{\hbar\omega}{\exp(\frac{\hbar\omega}{k_B T}) - 1}$$

and

$$\langle x \rangle = \frac{3b}{4a^2} \frac{\hbar\omega}{\exp(\frac{\hbar\omega}{k_B T}) - 1}. \tag{3.79}$$

Eq 3.79 reduces to Eq 3.76 at high temperatures. (At high T, $\hbar\omega/k_BT<<1$ and $\exp(\hbar\omega/k_BT)\fallingdotseq 1+(\hbar\omega/k_BT)$.)

- At low temperatures, however, Eq 3.79 can be written as,

$$\langle x\rangle=\frac{3b}{4a^2}\hbar\omega\cdot\exp(-\frac{\hbar\omega}{k_BT})$$

and the thermal expansion coefficient becomes,

$$\alpha=\frac{\partial\langle x\rangle}{\partial T}=\frac{3bk_B}{4a^2}(\frac{\hbar\omega}{k_BT})^2\exp(-\frac{\hbar\omega}{k_BT}) \qquad (3.80)$$

which decreases and eventually vanishes as temperature decreases to absolute zero.

Chapter Four

Electrons in Metal

- Energy Band Theory

4.1 Classical Theory of Free Electron

- There is no doubt about the existence of free electrons in metals. Free electrons are believed to do major role in atomic bonding, electric and thermal conduction and scattering electromagnetic waves of metallic materials. For understanding electromagnetic properties of solids, qualitatively and quantitatively, it is important to know the states of free electrons.

- In classical theory, the valence electrons in a metal are regarded as non-interacting particles like the ideal gas molecules. Free electrons are 'free' in this model, which means that free electrons hold kinetic energy only. Free electrons move freely except for collisions with ions, through which the vectorial average of momentum remains zero at equilibrium.

- Some properties of metals are explained rather satisfactorily by the classical theory though many important phenomena are not.

4.1.1 Electric conduction - Ohm's law

- The electrical conduction is dealt as a transport phenomenon here. In general, transport process can be expressed in the form of,

$$Q=t\nabla u. \tag{4.1}$$

Examples are given below.

Subject	Q	t	u
Mass	Particle flux	D : Diffusivity	n : Concentration
Energy	Heat flux	κ : Thermal conductivity	T : Temperature
Momentum	Momentum flux	η : Viscosity	v : Velocity
Charge	Current flux	σ : Conductivity	V : Voltage

- Now, the electric current flux(density) is written as,

$$j = \sigma \nabla V = -\sigma E . \qquad (4.2)$$

If σ is independent of E, we say that Ohm's law is obeyed.

- Using the free electron model, we can write

$$j = nev_D \qquad (4.3)$$

where n is number of free electrons per unit volume, e the electronic charge and v_D the so-called *drift velocity.*

- The drift velocity is the average velocity acquired by the electrons under an electric field. We know that the mean velocity of the electron without an applied field is zero because of random movements.

- The drift velocity of an individual electron becomes zero upon a collision with an atom(ion) since the scattering is isotropic. This means that the collision with ion acts as a retarding force on the electron giving a deceleration of v_D/τ. The relaxation time, τ, is the mean duration between two successive collisions. Then we can write an equation of motion for the electrons as,

$$m \cdot \frac{dv_D}{dt} = eE - m\frac{v_D}{\tau}.$$

Or,

$$m\left(\frac{dv_D}{dt} + \frac{v_D}{\tau}\right) = eE \; . \tag{4.4}$$

- In equilibrium, $dv_D/dt=0$ and we get

$$v_D = \frac{eE}{m} \cdot \tau \tag{4.5}$$

and

$$j = ne\frac{eE}{m}\tau = \frac{ne^2\tau}{m} \cdot E = \sigma \cdot E. \tag{4.6}$$

This is the Ohm's law. It is seen that

$$\sigma = \frac{ne^2\tau}{m} = \frac{1}{\rho} \tag{4.7}$$

where ρ is called the resistivity.

4.1.2 Thermal conduction

- In metals free electrons play an important role in heat transportation as well as phonons. In fact, the free electron is the major heat carrier. We can formulate the thermal conduction by the free electron using the ideal gas model.

- If there is a temperature gradient in a metal, heat will flow from higher-temperature point to lower-temperature point since the average kinetic energy of an electron is proportional to T ($3/2k_BT$ for classical model). The electrons at the hotter end($T=T_0+\triangle T$) have more energy than those at colder end($T=T_0$) by an average amount of $3/2k_B\triangle T$.

- This energy difference acts as the potential energy difference and gives rise to a driving force of thermal diffusion, $F=-(\partial U/\partial r)$. By this force the electrons drift toward to colder part with a drift velocity, v_D. Then net flow of energy per unit time and unit cross section will be expressed as

$$j_{hrat} = n \cdot q \cdot v_D \tag{4.8}$$

where q is the average kinetic energy of an electron, $3/2k_BT$.

- the drift velocity is nullified by collisions with ions, just like in the case of electrical conduction. Then we have an equation of motion for the electrons as,

$$m\frac{dv_D}{dt} = \frac{3}{2}k_B \nabla T - m\frac{v_D}{\tau} \tag{4.9}$$

and in equilibrium, $\frac{dv_D}{dt} = 0$, then we get,

$$v_D = \frac{\tau}{m} \cdot \frac{3}{2}k_B \nabla T \,. \tag{4.10}$$

- Eq 4.8 becomes

$$j_{heat} = n(\frac{3}{2}k_B)^2 T \cdot \frac{\tau}{m} \nabla T \tag{4.11}$$

and we get an expression for the thermal conductivity as,

$$x = (\frac{3}{2}k_B)^2 \cdot \frac{n\tau}{m} \cdot T \,. \tag{4.12}$$

- From Eq's (4.7) and (4.12), the following relation is obtained.

$$\frac{1}{T} \cdot \frac{\chi}{\sigma} = (\frac{3}{2} \cdot \frac{k_B}{e})^2 = constant. \tag{4.13}$$

Eq (4.13) is saying that the ratio, $(1/T)(\kappa/\sigma)$, has the same value for all metals and for all temperatures. This is known as the Wiedemann-Franz law and proved approximately true for many metals.

4.1.3 Failures of the classical free electron model

- The classical free electron theory which deal the electron as the ideal gas seems to explain the conduction phenomena, electrical and thermal, satisfactorily. But the theory faces crucial failures in explaining very small contribution of the free electron to the specific heat, C_v, and the paramagnetic susceptibility, χ.

- It is well known that the specific heat of an ideal gas is 3/2R per mole and the specific heat of a solid at elevated temperatures is 3R per mole which is due to lattice vibrations. Then it is natural to expect that the specific heat of a solid metal should be 3/2R+3R=9/2R due to the 'ideal gas' of electron and the lattice vibration. In practice, however, the specific heat of a metal remains 3R with very small increase with T. That is,

$$C_v = 3R + \gamma TR, \quad (\gamma \sim 10^{-4}) \tag{4.14}$$

for a metal.

- The paramagnetic susceptibility of a group of non-interacting magnetic moments can be written as

$$\chi = \frac{N\mu^2}{3k_B T}$$

where μ is the magnitude of the individual magnetic moment. We know that each electron has a magnetic moment of $1\mu_B$. If the free electron is an ideal gas, then the electronic contribution to the susceptibility should hold a value of,

$$\chi = \frac{N\mu_B^2}{3k_B T} \ .$$

In practice, the susceptibility shows values of,

$$\chi \simeq (10^{-4}T) \cdot \frac{N\mu_B^2}{3k_BT} \tag{4.15}$$

which is much smaller than expected and, furthermore, virtually independent of temperature.

4.2 Quantum Theory of Free Electron

4.2.1 Quantization of electronic states

- We know that electrons are subject to the quantum mechanics rather than to the classical mechanics. Ionic potentials assumed fixed, the steady-state Schrődinger equation can be adopted for the free electrons.

$$\left(-\frac{\hbar^2}{2m}\nabla^2 + V\right)\Psi = E\Psi\ . \tag{4.16}$$

Here, the first term in the left hand side represents the kinetic energy while V and E represent potential and total energies, respectively. Ψ, the wave function, relates with the probability to find the electron as $|\Psi|^2$ is the probability to find the electron at the given position and time. And so, Ψ is normalized by the following relation which means that if a particle exists somewhere in the universe the probability to find the particle in the universe is unity or 100%.

$$\int_{-\infty}^{\infty} |\Psi|^2 dv = 1. \tag{4.17}$$

- For free electrons, ionic potentials might be neglected, that is, V=0. Then we get a simpler equation,

$$-\frac{\hbar^2}{2m} \nabla^2 \Psi = E\Psi . \tag{4.18}$$

- For simplicity, one dimensional case is dealt with first.

$$-\frac{\hbar^2}{2m} \frac{\partial^2 \Psi}{\partial x^2} = E\Psi . \tag{4.19}$$

This differential equation has a general solution that,

$$\Psi = \Psi_0 e^{ikx} . \tag{4.20}$$

- Putting Eq (4.20) into Eq (4.19), we get,

$$E = \frac{\hbar^2 k^2}{2m} \quad . \tag{4.21}$$

- At this stage we do not see any quantization. Quantization comes from the confinement of a particle, which gives the boundary conditions. Let us consider a one dimensional metal piece of length L, and adopt the periodic boundary condition as,

$$\Psi(L+x) = \Psi(x) \quad . \tag{4.22}$$

Then, from Eq (4.20) and Eq (4.21) it is seen that,

$$e^{ikL} = 1$$

or,

$$k = \frac{2\pi}{L} \cdot n \quad , \quad (n = \pm 1, \pm 2 \cdots) \quad . \tag{4.23}$$

- Now it is seen that k-vector is quantized. Momentum and energy are also quantized as,

$$p_n = \hbar k_n$$

and

$$E_n = \frac{\hbar^2 k_n^2}{2m} = \frac{n^2 \hbar^2}{2m}\left(\frac{2\pi}{L}\right)^2 . \tag{4.24}$$

- This argument can be extended to three dimensional space. That is, for a cubic specimen with edge of length L, we can get,

$$k_x = n_x \frac{2\pi}{L}, \qquad n_x = 0, \pm 1, \pm 2, \cdots$$

$$k_y = n_y \frac{2\pi}{L}, \qquad n_y = 0, \pm 1, \pm 2, \cdots$$

$$k_z = n_z \frac{2\pi}{L}, \qquad n_z = 0, \pm 1, \pm 2, \cdots$$

- The momentum is quantized as,

$$\vec{p} = p_x \vec{i} + p_y \vec{j} + p_z \vec{k},$$

$$p_x = \hbar k_x, \quad p_y = \hbar k_y, \quad p_z = \hbar k_z$$

and

$$p^2 = \hbar^2 (k_x^2 + k_y^2 + k_z^2) = \hbar^2 \left(\frac{2\pi}{L}\right)^2 (n_x^2 + n_y^2 + n_z^2) . \tag{4.25}$$

- Energies are also quantized.

$$E=\frac{\hbar^2 k^2}{2m}=\frac{\hbar^2}{2m}(k_x^2+k_y^2+k_z^2)$$

$$=\frac{\hbar^2}{2m}\left(\frac{2\pi}{L}\right)^2(n_x^2+n_y^2+n_z^2)\,. \tag{4.26}$$

- We can say that a set of quantum numbers, (n_x, n_y, n_z), defines the momentum and energy. Since electrons have intrinsic spin, a set of four quantum numbers, (n_x, n_y, n_z, m_s) designates the state of an electron.

4.2.2 Fermi energy and Fermi surface

- Electrons obey the Pauli Exclusion Principle and one quantum state can accomodate one electron only. Every electrons try to occupy a state with energy as low as possible. This means that, if there is no thermal excitation, the electronic states will be occupied successively from the bottom without any vacancy until there is no free electron left. The highest energy among occupied states in this case (at T=0K) is called the Fermi energy, E_F. So, the Fermi energy may be expressed in the following way.

$$\int_0^{E_F} g(E)dE = N \tag{4.27}$$

where N is total number of the free electron and g(E) is the *density of states*. g(E)dE means the number of states with energies in between E and E+dE. We will be back on this matter later and for the time being the Fermi energy is obtained through a simpler method.

- It is seen from Eq (4.26) that the smaller k value is, the lower the energy is. We may say, therefore, that, in k-space, states are occupied in succession from those near the origin in the form of sphere.(Fig 4.1)

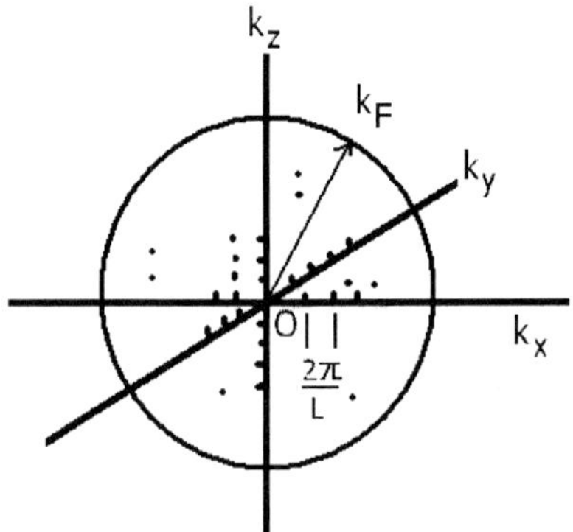

Fig 4.1 All the states inside the sphere of radius k_F are occupied at T=0K while states with $k>k_F$ are left unoccupied.

- The Fermi energy is then given by,

$$E_F = \frac{\hbar^2 k_F^2}{2m} \quad . \tag{4.28}$$

Here, k_F can be obtained from the relation that

$$N = \frac{volume\ of\ sphere\ (r = k_F)}{volume\ per\ a\ state} \times 2\ (spin\ up\ \text{and}\ down)$$

$$= \frac{\frac{4\pi}{3} k_F^3}{(\frac{2\pi}{L})^3} \times 2 = \frac{V}{3\pi^2} k_F^3$$

or

$$k_F = (3\pi^2 \frac{N}{V})^{\frac{1}{3}} = (3\pi^2 n)^{\frac{1}{3}} \tag{4.29}$$

where n is the number of free electrons per unit volume. And,

$$E_F = \frac{\hbar^2 k_F^2}{2m} = \frac{\hbar^2}{2m}(3\pi^2 n)^{\frac{2}{3}} \quad . \tag{4.30}$$

For an example, we can calculate E_F for copper.

$$n = \frac{\rho N_A}{M} = \frac{8.96 \times 10^3 \times 6.02 \times 10^{23}}{63.54 \times 10^{-3}} = 0.85 \times 10^{29} / m^3$$

$$E_F = \frac{(1.054\times 10^{-34})^2}{2\times 9.1\times 10^{-31}}(3\pi^2 n)^{\frac{2}{3}} = 1.13\times 10^{-18}J = 7eV .$$

- As seen in Fig 4.1, all the states with the Fermi energy lie on the surface of a sphere with a radius $k=k_F$ in the k-space. The surface is called the *Fermi surface.* In this free electron model, the Fermi surface has a form of sphere. We will see, however, the Fermi surface may show more complicated form in practice. In general, the Fermi surface is defined as a surface in the k-space, states on which have the Fermi energy.

4.2.3 Density of states

- In Eq 4.27, the density of states was already defined as number of states with a certain energy E in a body(per unit volume of the specimen, sometimes). Since intervals between energies of adjacent states is so small(order of Planck's constant), energies might be conveniently regarded as continuous rather than as discrete. In this circumstance, the density of states can be defined as g(E) when g(E)dE is the number of states whose energies lie in between E and E+dE.

- The density of states for the free electron model is obtained straightforwardly from the k-space diagram.(Fig 4.2) First, g(k)dk which is the number of states within the shaded thin shell in the figure is obtained and k and dk are converted into E and dE from the relation of Eq (4.28).

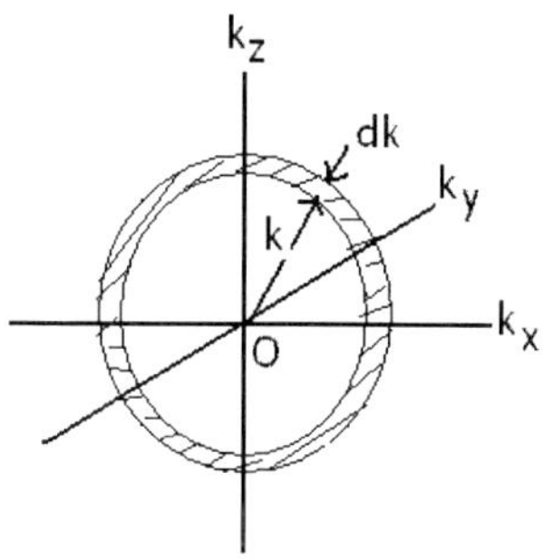

Fig 4.2 States in the shaded volume have energies in between E(k) and E(k)+dE.

- Now, the volume of the shell of radius k and of thickness dk (shaded part in Fig 4.2), is $4\pi k^2 \times dk$ and the volume per a state in k-space is $(2\pi/L)^3$. Then we get,

$$g(k)dk = \frac{4\pi k^2 dk}{(\frac{2\pi}{L})^3} \times 2(\pm spin)$$

$$=\frac{L^3k^2dk}{\pi^2}\left(=\frac{k^2dk}{\pi^2}(per\ unit\ volume)\right). \qquad (4.31)$$

- From Eq 4.28, we know that,

$$k^2=\frac{2mE}{\hbar^2} \qquad \text{and} \qquad 2kdk=\frac{2m}{\hbar^2}dE\ .$$

And we obtain,

$$g(E)dE=\frac{1}{\pi^2}\cdot\frac{m\sqrt{2m}}{\hbar^3}\cdot\sqrt{E}dE \qquad (4.32)$$

or,

$$g(E)=\frac{\sqrt{2}m^{\frac{3}{2}}}{\pi^2\hbar^3}\cdot E^{\frac{1}{2}}=\frac{1}{2\pi^2}\left(\frac{2m}{\hbar^2}\right)^{\frac{3}{2}}\cdot E^{\frac{1}{2}}. \qquad (4.33)$$

- Fig 4.3(a) shows a typical density-of-states(DOS) curve. The probability that a state is occupied by an electron, is decided by the Fermi-Dirac distribution function, f(E). (Fig 4.3(b))

$$f(E)=\frac{1}{\exp(\frac{E-E_F}{k_BT})+1} \tag{4.34}$$

- The number of electrons whose energies are in between E and E+dE is then given by,

$$N(E)dE=g(E)f(E)dE \tag{4.35}$$

and shows a typical N(E)-E diagram as in Fig 4.3(c).

- It is seen that difference between N(E) curves at 0K and T is some shift of energies of electrons near the Fermi surface while N(E) for electrons with energies far from the Fermi energy remain unchanged. This can be interpreted as electrons near the Fermi surface only can find vacant states to be excited into, while electrons far below the Fermi surface cannot find an empty state nearby and so cannot absorb energy at all.

- Comparing $E_F(T_F =\sim 10^{4\sim 5})$ and k_BT at room temperatures, we may say that electrons at the Fermi surface only contribute to energy exchange. For instance, specific heat of free electrons is due to not all the free electrons but electrons at the Fermi surface, $N(E_F)$, only. The specific heat due to free electrons, as a result, should be 10^{-4} of 3/2R that is expected when all the free electrons contribute to the specific heat.

- An electron has a magnetic moment of a Bohr magneton, μ_B, from its spin and in an applied magnetic field, H, its energy is increased by $\mu_B H$ when its moment orients oppositely to the field and then can be lowered as spin flips to the field direction. If all the free electrons execute such spin flips, fairly large susceptibility will be realized. But $\mu_B H$(~10^{-4} eV)in laboratory fields is much too small comparing with E_F(~ 10 eV) and we can understand only electrons near Fermi surface contribute to the paramagnetic susceptibility. This gives an answer to the failure of the classical free electron theory.

(a)

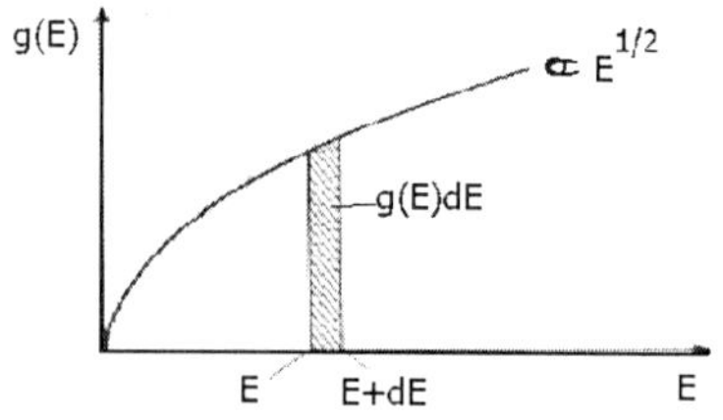

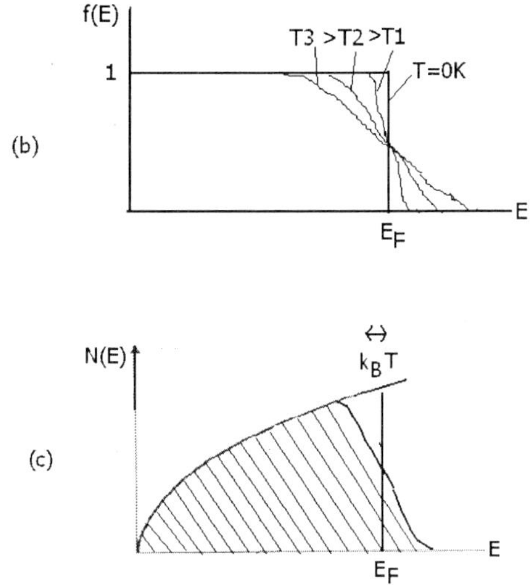

Fig 4.3 The density-of-states(DOS) curve of the free electron(a) and the Fermi-Dirac distribution (b) and the number of electrons, N(E) (c).

4.2.4 Failures of the simple quantum theory of free electron

- We have seen that energy exchange in ordinary level is executed by electrons at the Fermi surface only while most free electrons at states below E_F remain not participating. So the number of electrons at the Fermi surface, $N(E_F)$, is very significant in many physical properties of a solid.

- The above free electron theory gives an expression for $N(E_F)$. That is, from Eq's (4.30) and (4.33), we get,

$$N(E_F) = \frac{m}{\pi^2 \hbar^2}(3\pi^2 n)^{\frac{1}{3}} \propto n^{\frac{1}{3}} \qquad (4.36)$$

where n is the density of free electrons.

- At this point, two simple questions arise ; One is why electric conductivities are so different from metal to metal. The other is why Hall coefficients are so different from metal to metal and even positive in some metals. These questions cannot be answered by the simple quantum theory of free electron, since both electric conductivity and Hall coefficient depend on the density of free electrons which is not drastically different from metal to metal.

- Such difficulties come from the basic assumption that free electrons are 'really free'. But in practice electrons are more or less subject to electrostatic forces due to lattice of positive ions. When the electron-lattice interaction is taken into account, we meet the band theory.

- Before going further, the Hall effect is briefly explained. When a magnetic field H is applied on a current carrying strip perpendicular to the plane of the strip, charge carriers are exerted a force by the magnetic field to one of the two edges. An electric field E is resulted across the strip, whose intensity is just to make exact balance between the magnetic force and the electrostatic force. (Fig 4.4)

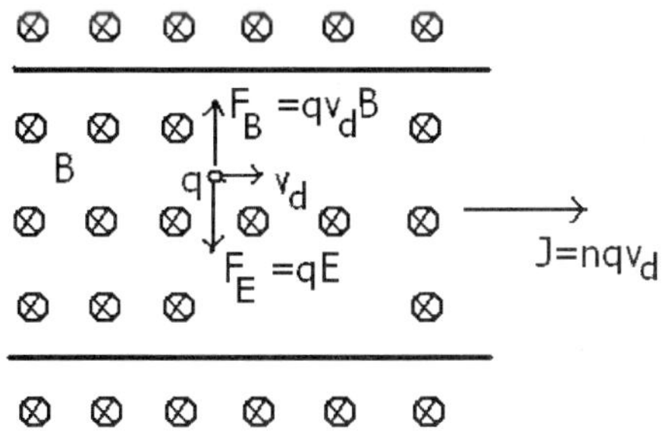

Fig 4.4 Magnetic force on the charge carriers is exactly canceled out by the electrostatic force due to the Hall field.

- When the charge carrier of the metal is of one kind(electron or hole, say), the Hall field is obtained from the following relation.

$$qE = qv_dB \tag{4.37}$$
$$\therefore E = v_dB = \frac{1}{nq} \cdot J \cdot B$$

Or,

$$\frac{E}{B} = \frac{1}{nq} \cdot J = R_H \cdot J\ . \tag{4.38}$$

- The coefficient, R_H (=1/nq), is called the Hall coefficient and holds positive value when q is positive(for the case of holes). This means that the Hall electric field directs as shown in Fig 4.4. In metals, since charge carriers are electrons, the Hall coefficient should be negative and the hall electric field directs toward upper side from the bottom side in the Fig 4.4. Most metals show negative Hall coefficient in practice. In some metals, Bi for instance, however, positive Hall coefficients appear, which cannot be understood on the free electron model.

4.3 Energy Band Theory

4.3.1 Nearly free electron model

- There are a few lines of theory to understand the formation of energy bands of free electrons. They may be grouped into two distinctive categories; the nearly free electron (NFE) model is one and the linear combination of atomic orbitals (LCAO) is the other. The latter is rather 'theoretical' while the former is phenomenological in a way. The NFE model is, therefore, dealt first here for approaching to the concept of band theory. Later the LCAO method will be explained in short.

- In the free electron theory, electrons are assumed unaffected by the ions(or lattice) and that's why the potential energy is kept uniform throughout the crystal(V=0). In the NFE model. it is also assumed same for most free electrons. But some electrons which satisfy a certain condition are strongly affected by the lattice. The condition is the Bragg condition and the result is reflection of the wave of electrons. In other words, electrons at the Brillouin zone boundary experience the Bragg reflection. And as the consequency appear energy bands.

- For simplicity, the crystal is taken as one dimensional lattice with the interatomic distance (lattice parameter)of a and length of L. 'Most' electrons may be expressed by wave functions as,

$$\psi_m = \psi_0 \cdot e^{-i(\omega t - kx)}, \quad (k = \frac{2\pi}{L} \cdot n) . \qquad (4.39)$$

- When the Bragg condition is met, i.e., k=nπ/a, the electronic wave is reflected by the 'plane' of the lattice. The reflected waves interfere with the on-going waves to form standing waves.

$$\begin{aligned} \psi_{on} &= \psi_0 \cdot e^{-i(\omega t - kx)} \quad && on-going \quad electrons \\ \psi_{re} &= \psi_0 \cdot e^{-i(\omega t + kx)} \quad && reflected \quad electrons \end{aligned} . \qquad (4.40)$$

The standing waves form through linear combination of on-going and reflected waves. The linear combination can be made in two ways; addition and subtraction, resulting in two distinctive standing waves.

$$\psi_{st}^{\pm} = \psi_{on} \pm \psi_{re}.$$

That is,

$$\begin{aligned}\psi_{st}^{+} &= \psi_{on} + \psi_{re} = \psi_0 \cdot e^{-i\omega t} \cdot [e^{ikx} + e^{-ikx}] \\ &= 2\psi_0 \cos kx \cdot e^{-i\omega t} \\ &= 2\psi_0 \cos(\frac{\pi}{a}x) \cdot e^{-i\omega t} \qquad (taking\ k\ as\ \frac{\pi}{a})\end{aligned} \tag{4.41}$$

Similarly,

$$\psi_{st}^{-} = \psi_{on} - \psi_{re} = 2i\psi_0 \sin(\frac{\pi}{a}x) \cdot e^{-i\omega t}. \tag{4.42}$$

- Probabilities of finding for the three kinds of electrons, Ψ_m, Ψ^+_{st} and Ψ^-_{st}, say, are quite different to one another. That is,

$$|\psi_m|^2 = \psi_0^2, \quad |\psi_{st}^{+}|^2 = 4\psi_0^2 \cos^2(\frac{\pi}{a}x),$$
$$|\psi_{st}^{-}|^2 = 4\psi_0^2 \sin^2(\frac{\pi}{a}x). \tag{4.43}$$

It is seen that most free electrons have a uniform probability of finding through the crystal while electrons at the Brillouin zone boundary show fluctuating probabilities. Electrons represented by Ψ^{+}_{st} show highest probability near ions (x=na) and least probability far from ions (x=(n+1/2)a). Electrons represented by Ψ^{-}_{st} show probabilities of finding in the other way round.(Fig 4.5)

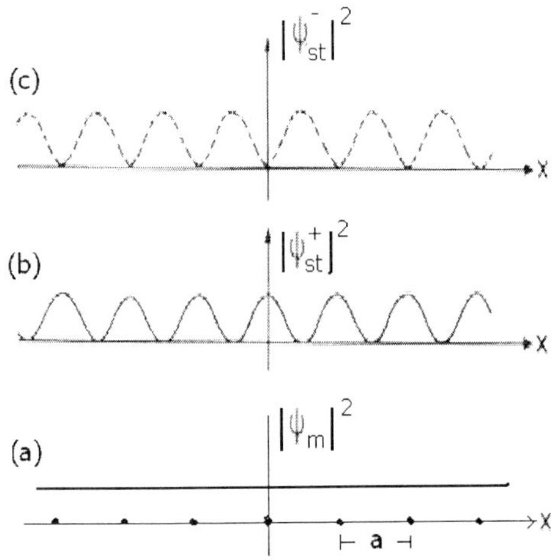

Fig 4.5 Probabilities of finding of 'most' free electrons (a) and electrons at the Brillouin zone boundary(b, c).

- Since all the 'most' free electrons which are away from the Brillouin zone have homogeneous probabilities to be near or far from ions no matter what the k-value is, the electrostatic potential energy should be same for all the electrons and therefore we set the potential energy as zero (V=0) for the Schroedinger equation (Eq 4.18) to get an expression for the energy, $E = \frac{\hbar^2 k^2}{2m}$ (Eq 4.21). Electrons at the zone boundary represented by the standing waves, Ψ^+_{st} and Ψ^-_{st}, however, are not the case. Electrons which are more probable to be near ions than 'free' electrons should have lower electrostatic potential energy. And the potential energy might be expressed in an average as $-V_G$ which is the Fourier integral over space, instead of V=0. For electrons which are in between ions, the potential energy might be taken as $+V_G$. Then the total energy, kinetic and potential, should be approximated as

$$E = \frac{\hbar^2 k^2}{2m} \pm V_G \tag{4.44}$$

at the Brillouin zone. E-k curve, consequently, deviates from a parabolic one (Eq 4.21) near the Brillouin zone boundaries, resulting in energy gaps, $E_g \sim 2V_G$. (Fig 4.6)

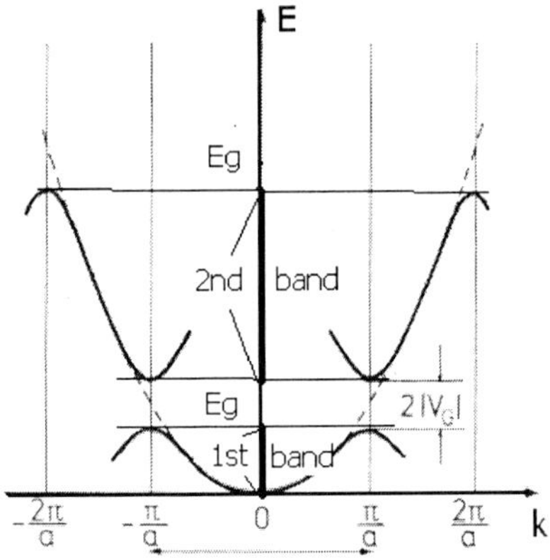

Fig 4.6 Energy parabolic of free electrons is splitted at the Brillouin zone boundaries to form bundles of energy levels, energy bands, say.

4.3.2 Density of states

- We have seen that DOS of free electron is proportional to $E^{1/2}$, that is,(Eq 4.33)

$$g(E) \propto \sqrt{E}.$$

This relation was obtained from the assumption of spherical equi-energy shell in k-space which is true for free electrons (Eq's 4.31, 4.32). Then, what will happen to energy contours in the NFE model?

- In the NFE model, energy contours remain spherical when k's are remote from the Brillouin zone boundary and electrons behave like free electrons. DOS in such region follows the $E^{1/2}$ curve. As k's approach the zone boundary, energy contours start to deviate from spherical form bulging toward the zone boundary (Fig 4.7), since energies of electrons near the zone boundary are smaller than those of as-free electrons as seen in Fig 4.6.

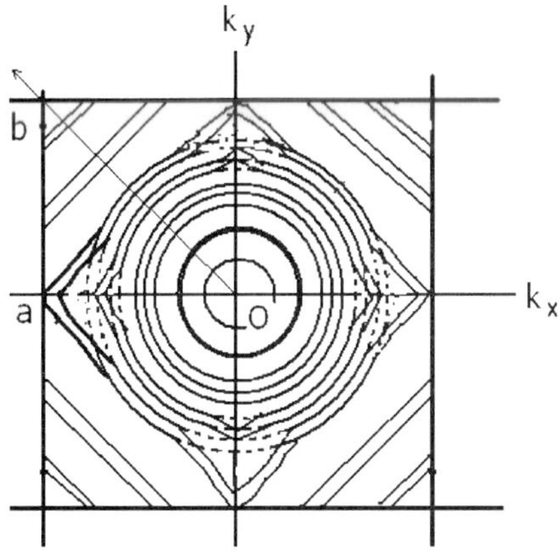

Fig 4.7 Energy contours in the 1st Brillouin zone for a simple 'square' crystal. As k increases, contours bulge toward the zone boundary, touch the boundary and break into pieces.

- The DOS is the number of states with energies in between E and E+dE, which is proportional to the volume of equi-energy shell in Fig 4.6. When the contour bulges, the surface becomes larger than sphere and so does the volume of the shell of thickness of δE. This means that DOS at near the Brillouin zone boundary becomes larger than that of as-free electrons. Such tendency continues until the energy whose contour touches the zone boundary. Over that energy, contours split and the volume of the shell decreases rapidly with E, resulting in a sharp peak in the DOS curve. (Fig 4.8a)

- At the point 'a' in Fig 4.7, there is an energy gap E_g between energies of states of 1st and 2nd Brillouin zone, E_{a1} and E_{a2}. Energy of the state 'b', E_b, is obviously higher than E_{a1}. If $E_{a2} > E_b$, 1st and 2nd energy bands are separated by a resultant energy gap, $E_g = E_{a2} - E_b$. (Fig 4.8b) If $E_{a2} < E_b$, however, 1st and 2nd bands overlap. This is called the *band overlapping*. (Fig 4.8c)

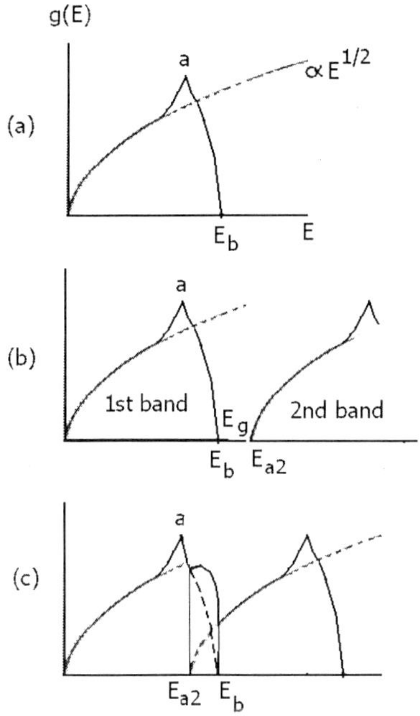

Fig 4.8 Density of states curves for 1st Brillouin zone (a), non-overlapping bands (b) and overlapping bands (c).

4.3.3 Metal, insulator and semiconductor

- Number of electrons which one energy band can accomodate is two per a primitive unit cell. This can be obtained straightforwardly from the fact that number of states per a Brillouin zone is two per unit cell. That is,

$$\frac{volume\ of\ a\ B.Z.}{volme\ of\ a\ k-state}\times 2(spin)=\frac{(\frac{2\pi}{L})^3}{(\frac{2\pi}{a})^3}\times 2=\frac{L^3}{a^3}\times 2$$

$$=2n\ (n: number\ of\ unit\ cells\ in\ the\ body)$$

$$=2\ per\ a\ unit\ cell \tag{4.45}$$

- If there is one free electrons per a unit cell, the 1st band is half filled and E_F lies at the middle of the band. And lots of unoccupied states exist adjacent to the uppermost electrons. This makes the material conductor or metal. For an example, Cu crystalizes in fcc with four atoms per a cell but the primitive cell is rhombohedral with one atom per a cell. We know that a Cu atom provide one free electron. So there is one electron in a cell and Cu should show good electric conductivity. In fact, most of elements with odd number of free electrons per an atom are metals.

- Bi with 5 electrons at the outer orbit, however, has two atoms in a unit cell and ten free electrons all together in a unit cell which would form five completely filled bands with an energy gap on the top of the 5th band. This means no electron can participate in energy exchange. Or there should be no charge carrying electrons and Bi should be an insulator. But in practice, 5th and 6th bands of Bi overlap slightly to result in 'holes' at the top of 5th band and simultaneously electrons at the bottom of 6th band. These carriers make Bi a 'poor' conductor instead of an insulator.

- On the other hand, elements with even number of free electrons would have completely filled bands with energy gap between filled and empty bands. This picture describes a typical band structure of an insulator. But in many of such elements, band overlapping occurs to make those metal.

- If energy gap of an insulator is not very large, ~1 eV, say, electrons at the top of the filled band can jump up to the upper empty band with help of thermal energy (~0.025 eV) with finite probabilities as temperature goes up to room temperature. In such materials, there are some holes in the filled band (valence band) and some electrons in the empty band (conduction band). These electrons and holes do their role as charge carrier but because of small number give lower conductivity comparing with metals. Such elements, Si and Ge for example, are called semiconductor.

4.3.4 Effective mass

- If electrons are 'free', their mass will be naturally m, electronic mass. When a 'free' electron is exerted on by a force, F, it will be accelerated. That is,

$$a = \frac{F}{m} \quad . \tag{4.46}$$

- As seen above, some electrons are not 'free' but affected by the lattice potential. For an example, let's consider an electron whose k-value is just below k=π/a in Fig 4.6 . If a force exerts on this electron in +x direction and does a work to increase the energy of this electron, then k-value of the electron will increase and touches the Brillouin zone. The electron will be reflected into -x direction. If force and resultant acceleration only are taken into account, we may say that a force in +x direction results in an acceleration in -x direction on this electron. Or, from the force law of Newton (m=F/a), it can be said that the electron has a negative mass. We know such an apparent mass is not the real mass. The apparent mass is called the *effective mass*. Furthermore, it can be easily understood that the magnitude of effective mass of the above case is far smaller than the real electronic mass. In some cases, effective mass is much heavier than electronic mass.

- A simple expression of the effective mass, m^*, is obtained from the following relations;

$$E = \hbar\omega, \quad P = \hbar k, \quad F = \frac{dP}{dt} = \hbar\frac{dk}{dt},$$

$$v_g = \frac{d\omega}{dk} = \frac{1}{\hbar}\cdot\frac{dE}{dk}. \tag{4.47}$$

Acceleration is,

$$a = \frac{dv_g}{dt} = \frac{1}{\hbar}\cdot\frac{d^2E}{dk\cdot dt} = \frac{1}{\hbar}\cdot\frac{d^2E}{dk^2}\cdot\frac{dk}{dt} = \frac{1}{\hbar^2}\cdot\frac{d^2E}{dk^2}\cdot F \tag{4.48}$$

Then the effective mass is expressed as,

$$m^{*} = \frac{F}{a} = \frac{\hbar^{2}}{\left(\frac{d^{2}E}{dk^{2}}\right)} \quad . \tag{4.49}$$

In three dimensional crystal, the effective mass has a tensor form as,

$$\frac{1}{m_{ij}^{*}} = \frac{1}{\hbar^{2}} \cdot \frac{\partial^{2}E}{\partial k_{i} \cdot \partial k_{j}} . \tag{4.50}$$

- Effective masses of electrons at different parts of Brillouin zone for a simple one dimensional case are analysed below.

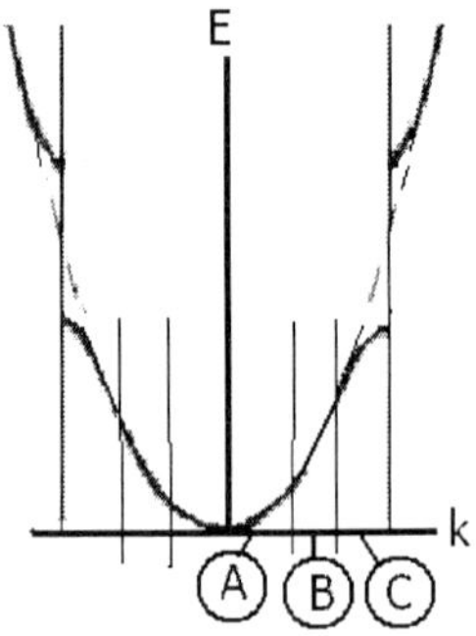

Fig 4.9 Brillouin zone is divided into three regions for qualitative analysis of effective mass.

Region Ⓐ : Electrons behave as free electrons and the dispersion curve (E-k curve) is almost same as that of free electron. That is,

$$E=\frac{\hbar^2 k^2}{2m} \quad \text{and} \quad \frac{\partial^2 E}{\partial k^2}=\frac{\hbar^2}{m} .$$

$$\therefore \; m^*=\frac{\hbar^2}{\left(\frac{\hbar^2}{m}\right)}=m .$$

So, effective mass of electrons in region A is same as the electronic mass.

Region Ⓑ : The slope of the dispersion curve at this region changes with k very little. That is, $\frac{\partial^2 E}{\partial k^2}$ has very small value. Therefore, m*'s have large magnitudes in this region.

Region Ⓒ : The slope of the dispersion curve decreases as k increases. This gives $\frac{\partial^2 E}{\partial k^2}<0$ and negative effective masses. The rate of change of the slope becomes bigger as k approaches the Brillouin zone boundary so that small and negative effective masses appear near the zone boundary.

4.3.5 Holes

- In a band, the electronic behaviors can be deduced from the behaviors of the 'vacant' state. (See Fig 4.10)

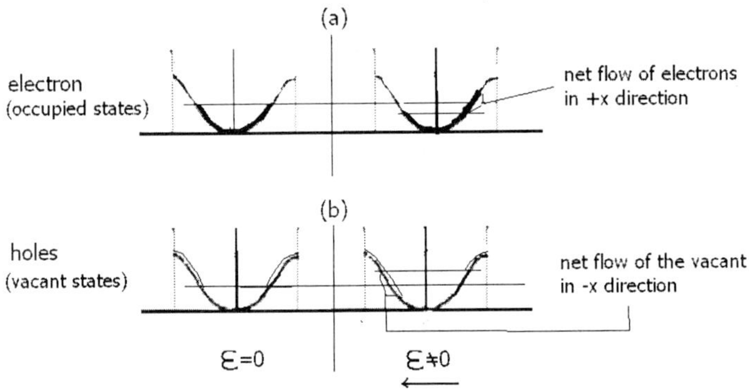

Fig 4.10 Electronic states shifted by a field(a) can be expressed by shift of vacant states(b).

These 'net' flows of electron and the vacant represent the same flow of electricity. This means that the 'charge' of the vacant is of opposite sign to that of electron, i.e., +e.

- The energy of the vacant increases as it moves down from the top of the band. The effective mass of the vacant is of same magnitude as that of corresponding electron with opposite sign. These vacant states are called *holes*. We can interpret physical phenomena of solids in terms of behaviors of holes instead of electronic behavior.

- When the number of holes is comparable with that of electrons, however, it is of no benefit to introduce holes in place of electrons. But when the number of holes is considerably smaller than electrons, it is convenient to deal with holes instead of electrons.

- For overlapping bands, if the overlapping is rather heavy, we don't need to introduce holes. When the overlapping is slight, however, it is easier to deal with electrons at the upper band and holes at the lower band.

- In intrinsic semiconductors, some electrons at the filled band(valence band) are excited into the upper empty band (conduction band). Electrical properties of semiconductors are well explained by the roles of electrons in conduction band and holes in the valence band.

- The Hall coefficient for the case that there exist electrons and also holes is given as,

$$R = \frac{R_e \sigma_e^2 + R_h \sigma_h^2}{(\sigma_e + \sigma_h)^2} \tag{4.51}$$

where,

$$R_e = -\frac{1}{n_e e}, \qquad R_h = \frac{1}{n_h e}$$

and

$$\sigma_e = \frac{n_e e^2 \tau_e}{m_e^*}, \qquad \sigma_h = \frac{n_h e^2 \tau_h}{m_h^*}.$$

4.3.6 Linear combination of atomic orbitals

- Free electrons belonged to individual constituent atoms initially before forming the group of free electrons. That is to say, when atoms are far away and independent from each other, electronic states are quantized within each atomic system to have sharply defined energies. The electronic states or orbitals are represented by wave functions of each atom.

- If two atoms are brought closer to each other, outer orbitals begin to overlap and electrons in the overlapping orbitals see different environments from those before overlapping, e.g., potentials from two nuclei instead of one each nucleus. Wave functions of the electrons, therefore, are perturbed to new wave functions to meet new environments. One of theoretical approaches to get new wave function is to start with the linear combination of each atomic orbitals(LCAO). Energy levels of 'new' orbitals become splitted into two from the sharp level before the overlapping. These splitted energy states now can accomodate the overlapping electrons from both atoms under the Pauli exclusion principle.

- In a solid, the perturbation is made over all the atoms and splitting of energy levels also arises into same number as atoms. Magnitude of the perturbation or width of energy splitting depends on the extent of overlapping, or on the interatomic distance. Fig 4.11 shows a schematic expression of energy level splitting for Si crystal. It is seen that, as interatomic distance decreases, energies of s-states and p-states intersect making overlapping range. The overlapped range of energy are forbidden to occupy and forms a energy gap between the lower occupied energy band(valence band) and the upper empty energy band(conduction band).

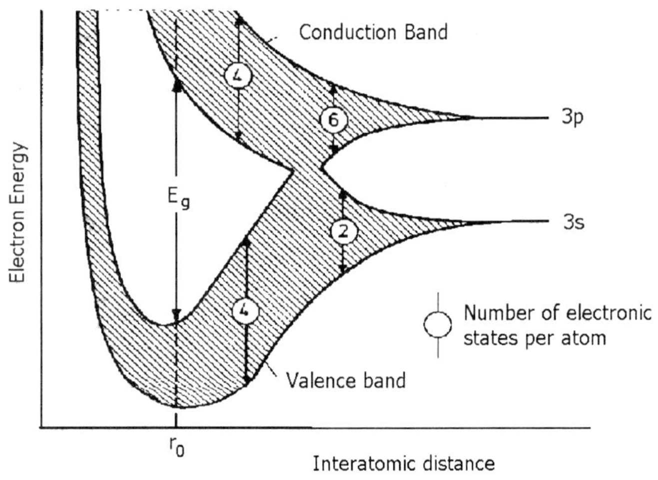

Fig 4.11 Schematic expression of energy bands as a function of interatomic distance. It is seen that s-and p-orbitals overlap to form two energy bands with an energy gap in between as the atomic separation decreases. In this range, Si with the diamond structure is known to have the sp3 hybrid orbitals.

- The LCAO type of approaches seem to have more generality than the NFE model and give a good base for the energy band theory for amorphous materials in particular. The NFE model, however, provides easier understanding of forming of energy bands and the Brillouin zone boundary effects. Anyway, the truth is one that energy bands and energy gaps exist, no matter which and how theory is explaining.